Lecture Notes in Bioengineering

Lecture Notes in Bioengineering (LNBE) publishes the latest developments in bioengineering. It covers a wide range of topics, including (but not limited to):

- Bio-inspired Technology & Biomimetics
- Biosensors
- Bionanomaterials
- Biomedical Instrumentation
- Biological Signal Processing
- Medical Robotics and Assistive Technology
- Computational Medicine, Computational Pharmacology and Computational Biology
- Personalized Medicine
- Data Analysis in Bioengineering
- Neuroengineering
- Bioengineering Ethics

Original research reported in proceedings and edited books are at the core of LNBE. Monographs presenting cutting-edge findings, new perspectives on classical fields or reviewing the state-of-the art in a certain subfield of bioengineering may exceptionally be considered for publication. Alternatively, they may be redirected to more specific book series. The series' target audience includes advanced level students, researchers, and industry professionals working at the forefront of their fields.

Indexed by SCOPUS, INSPEC, zbMATH, SCImago.

More information about this series at http://www.springer.com/series/11564

Kamalpreet Sandhu · Sunpreet Singh ·
Chander Prakash · Neeta Raj Sharma ·
Karupppasamy Subburaj
Editors

Emerging Applications of 3D Printing During CoVID 19 Pandemic

Springer

Editors
Kamalpreet Sandhu
Department of Product and Industrial Design
Lovely Professional University
Phagwara, Punjab, India

Chander Prakash
School of Mechanical Engineering
Lovely Professional University
Phagwara, Punjab, India

Karupppasamy Subburaj
Engineering Product Development Pillar
Singapore University of Technology and Design
Singapore, Singapore

Sunpreet Singh
National University of Singapore
Singapore, Singapore

Neeta Raj Sharma
School of Bioengineering and Biosciences
Lovely Professional Univeristy
Phagwara, Punjab, India

ISSN 2195-271X ISSN 2195-2728 (electronic)
Lecture Notes in Bioengineering
ISBN 978-981-33-6702-9 ISBN 978-981-33-6703-6 (eBook)
https://doi.org/10.1007/978-981-33-6703-6

This Springer imprint is published by the registered company Springer Nature Singapore Pte Ltd.
The registered company address is: 152 Beach Road, #21-01/04 Gateway East, Singapore 189721, Singapore

Preface

The book entitled *Emerging Application of 3D printing during COVID-19 Pandemic* presents various practical outbreaks of 3D printing technologies on developing different types of tools and gadgets to get prepared for fighting COVID-19. This book presents multidisciplinary aspects of the evolutionary growth of this exceptional technology, including social, medical, administration, and scientific. This book presents state-of-the-art applications of 3D printing technology including the development of PPE, ventilators, respiratory, and customized drugs. Moreover, a variety of research activities, at R&D centers, academic institutions, and commercial enterprises, are covered via incorporating research, review, technical notes, and short communications. Overall, it is believed that the combined efforts of the editorial team members and contributing authors will provide this book a huge attention across R&D, manufacturing, medical, and academic platforms.

Phagwara, India	Kamalpreet Sandhu
Singapore, Singapore	Sunpreet Singh
Phagwara, India	Chander Prakash
Phagwara, India	Neeta Raj Sharma
Singapore, Singapore	Karupppasamy Subburaj

Contents

Practical Frontline 3D Printing of Biomedical Equipment: From Design to Distribution—A North American Experience 1
Leonid Chepelev, Prashanth Ravi, and Frank J. Rybicki

The Role of Distributed Manufacturing and 3D Printing in Development of Personal Protective Equipment Against COVID-19 15
John Scott Frazer

Effectiveness of 3D Printing and Open-Source Technologies for Development of Ventilators, and Other Critical Care Technology in the Context of the COVID-19 Pandemic 35
John Scott Frazer

Artificial Intelligence in 3D Printing: A Revolution in Health Care 57
Aishwarya Banerjee, Haritha K. Haridas, Arunima SenGupta, and Neetu Jabalia

3D Printing Technology for Fighting COVID-19 Pandemic 81
Rohin Shyam, Pearlin Hameed, P. Suya Prem Anand, Loganathan Rangasamy, Arunkumar Palaniappan, and Geetha Manivasagam

Fostering Novel Materials and Subsisting Technologies for 3D Printing 111
Himangi Sood, Simran Kaur, and Ishani Sharma

COVID-19 Safe Clinical Dentistry: Applications of 3D Printing Technologies 125
Anoop Kapoor, Ishdeep Kaur, Arshdeep Kaur, Himanshu Deswal, Rajni Jain, Prabhjot Kaur, and Vishakha Grover

Developing Personal Protective Equipment Against COVID-19 153
Sagarika Bhattacharjee and Harmanpreet Singh

Use of Additive Manufacturing in the Battle Against COVID-19 167
Harish Kumar Banga, Rajesh Kumar, Parveen Kalra, R. M. Belokar, and Somya Saxena

3D Printing During COVID-19: Challenges and Possible Solutions 179
Jyotindra Narayan, Sanchit Jhunjhunwala, and Santosha K. Dwivedy

Covid-19 Success Stories of 3D Printing 193
Harmanpreet Singh, Sagarika Bhattacharjee, and Puneet Bawa

About the Editors

Kamalpreet Sandhu is an Assistant Professor in the Product and Industrial Design department at Lovely Professional University, Phagwara, Punjab, INDIA. His primary focus is on the design and development of footwear products and injuries prevention. He was done different projects in Podiatric Medicine at the Defence Institute of Physiology and Allied Sciences, DRDO, Delhi, i.e., Design and developed a new kind of orthosis for social needs and work resulted in a publication "Effect of Shod Walking on Plantar Pressure with Varying Insole". His area of research is Design Thinking, 3D printing and Ergonomics for Podiatric Medicine. He is also the editor of various books: Sustainability for 3D Printing, Revolutions in Product Design for Healthcare, Food Printing: 3D printing in Food Sector and 3D printing in Podiatric Medicine. He is also acting as an Editorial Review board member for the International Journal of Technology and Human Interaction (IJTHI), Advances in Science, Technology and Engineering Systems Journal (ASTESJ) and also a review editor for Frontiers in Manufacturing Technology section "Additive Processes". He has established a research collaboration with Prof. Karupppasamy Subburaj at Singapore University of Technology and Design (SUTD), SINGAPORE on Medical Device Design and Biomechanics.

Sunpreet Singh is researcher in NUS Nanoscience & Nanotechnology Initiative (NUSNNI). He has received Ph.D. in Mechanical Engineering from Guru Nanak Dev Engineering College, Ludhiana, India. His area of research is additive manufacturing and application of 3D printing for development of new biomaterials for clinical applications. He has contributed extensively in additive manufacturing literature with publications appearing in Journal of Manufacturing Processes, Composite Part: B, Rapid Prototyping Journal, Journal of Mechanical Science and Technology, Measurement, International Journal of Advance Manufacturing Technology, and Journal of Cleaner Production. He has authored more than 150+ research papers and book chapters. He is working in joint collaboration with Prof. Seeram Ramakrishna, NUS Nanoscience & Nanotechnology Initiative and Prof. Rupinder Singh, manufacturing research lab, GNDEC, Ludhiana. He is also editor of 3 books- "Current Trends in Bio-manufacturing", "3D Printing in Biomedical

Engineering", and "Biomaterials in Orthopaedics and Bone Regeneration - Design and Synthesis". He is also guest editor of several journals- special issue of "Functional Materials and Advanced Manufacturing", Facta Universitatis, series: Mechanical Engineering (Scopus Indexed), Materials Science Forum (Scopus Indexed), and special issue on "Metrology in Materials and Advanced Manufacturing", Measurement and Control (SCI indexed), Materials, Sustainibility (MDPI).

Chander Prakash is Associate Professor in the School of Mechanical Engineering, Lovely Professional University, Jalandhar, India. He has received Ph. D in mechanical engineering from Panjab University, Chandigarh, India. His areas of research are biomaterials, rapid prototyping & 3-D printing, advanced manufacturing, modeling, simulation, and optimization. He has more than 11 years of teaching experience and 6 years of research experience. He has contributed extensively to the world in the titanium and magnesium based implant literature with publications appearing in Surface and Coating Technology, Materials and Manufacturing Processes, Journal of Materials Engineering and Performance, Journal of Mechanical Science and Technology, Nanoscience and Nanotechnology Letters, Proceedings of the Institution of Mechanical Engineers, Part B: Journal of Engineering Manufacture. He has authored 60 research papers and 10 book chapters. He is also editor of 3 books: "Current Trends in Bio-manufacturing"; "3D Printing in Biomedical Engineering"; and "Biomaterials in Orthopaedics and Bone Regeneration - Design and Synthesis". He is also guest editor of 3 journals: special issue of "Functional Materials and Advanced Manufacturing", Facta Universitatis, Series: Mechanical Engineering (Scopus Indexed), Materials Science Forum (Scopus Indexed), and special issue on "Metrology in Materials and Advanced Manufacturing", Measurement and Control (SCI indexed).

Neeta Raj Sharma in Biochemistry from Jiwaji University, Gwalior, is currently leading the School of Bioengineering and Biosciences as Additional Dean in Lovely Professional University, Phagwara, Punjab. She is a visiting Professor at Birmingham City University, UK and is actively working in association with several esteemed universities in Canada like University of British Columbia, McGill University, Laval University and University of Victoria. With a vast multidisciplinary research experience of thirteen years in the field of Bioengineering and Biosciences, Prof. Sharma presently envisions extensive biochemistry and biotechnology based methods to develop diagnostic tools and employ bioanalytical instrumentation pertaining to apt diagnosis in the domain of biomedical engineering and other allied research sectors. She has been conferred with prestigious research awards from LPU twice. She is a fellow member of Association of Biotechnology and Pharmacy, India and also bears to her credential membership of prestigious societies like Indian Science Congress, Association of Indian Science Congress, Association of Microbiologists of India, Indo-US Collaboration of Engineering Education 2013, Fellow Member International Science Congress Association, Life membership of association for promotion of DNA Finger Printing and other DNA

Technologies. Till date, she has published more than 55 publications in peer-reviewed journals of high repute, 20 patents, 4 copyrights, 2 edited books and several articles in reputed magazines. She has been currently acting as reviewer for several Scopus-indexed journals of high standing and esteem and was appointed as guest editor for Scopus-indexed journal.

Karupppasamy Subburaj is an Assistant Professor in the Pillar of Engineering Product Development (EPD) at Singapore University of Technology and Design (SUTD). He leads an interdisciplinary research team to design and develop medical devices, assistive technologies, image-based quantitative biomarkers, and computing tools (machine learning, artificial intelligence) for diagnosing, monitoring, treating, and potentially preventing musculoskeletal disorders (osteoarthritis/osteoporosis) and disabilities by understanding bio-mechanical implications of those diseases and disabilities. He collaborates with physicians and clinical researchers from Tan Tock Seng Hospital (TTSH), Technical University of Munich (TUM), Singapore General Hospital (SGH), and Changi General Hospital (CGH) to combine research and technical expertise to address real-life clinical problems affecting Asia-Pacific and the World. Before joining SUTD, he did his postdoctoral work in the Musculoskeletal Quantitative Imaging Research (MQIR) laboratory at the University of California San Francisco (UCSF). At UCSF, he worked with a spectrum of clinicians (from radiology, orthopaedic surgeons, sports medicine, and physiotherapy/rehabilitation science) on characterizing magnetic resonance image (MRI) based bio-markers to understand the physiological and biochemical response of knee joint cartilage to physical exercise and acute loading. He has also developed and validated 3D modelling and quantification methods to study joint (hip/knee) loading patterns and contact kinematics in young healthy adults and patients with osteoarthritis. He received his PhD from the Indian Institute of Technology Bombay (IIT Bombay), India, in 2009. During his PhD, he collaborated with Tata Memorial Hospital (TMH), Mumbai, on developing a Surgery Planning System for Tumour Knee Reconstruction. He also worked with orthodontists from local hospitals in Mumbai, India, on designing and developing prostheses and surgical instruments/guides for reconstructing maxillofacial defects. After his PhD, he worked as a research specialist (surgery planning) in the Biomedical Engineering Technology incubation Centre (BETiC) at IIT Bombay, before moving to UCSF for his postdoctoral studies.

Practical Frontline 3D Printing of Biomedical Equipment: From Design to Distribution—A North American Experience

Leonid Chepelev, Prashanth Ravi, and Frank J. Rybicki

Abstract With its versatility, wide availability, and a worldwide active community of enthusiasts, scientists, engineers, and physicians, 3D printing has demonstrated practical value and potential in providing stopgap solutions to shortages of key equipment. Despite enthusiastic support for 3D printing to meet some equipment shortages, the effectiveness of practical implementation of such prototypes has been variable. In this work, we draw on the practical experiences of our groups in Canada and in the United States that used 3D printing for pandemic-related equipment shortages. We describe challenges and solutions for implementing and coordinating programs for 3D printing response in addressing shortages of personal protective equipment (PPE), specialized equipment for intubation and respiratory support, and development of simpler hardware to extend the lifecycle and applications of existing equipment.

Keywords COVID-19 · 3D printing · Personal protective equipment

1 Introduction

The sudden explosive growth in COVID-19 cases and the ensuing shutdown or slowdown in manufacturing and logistics operations worldwide, coupled with limited stockpiles of key medical equipment locally have created a perfect storm of supply shortages. The severity of equipment shortages was not homogeneous across North America and was driven by a combination of local case volumes, testing availability, local healthcare system capacity, and the uncertainty related to projected evolution of case volumes, among other factors. Both, the shortages of vital equipment and the potential for stopgap manufacturing using 3D printing were recognized by the United States Food and Drug Administration (FDA) and Health Canada, resulting

L. Chepelev (✉)
Department of Radiology, Stanford University, Stanford, CA 94305, USA
e-mail: chepelev@stanford.edu

P. Ravi · F. J. Rybicki
Department of Radiology, University of Cincinnati, Cincinnati, OH 45221, USA

K. Sandhu et al. (eds.), *Emerging Applications of 3D Printing During CoVID 19 Pandemic*, Lecture Notes in Bioengineering, https://doi.org/10.1007/978-981-33-6703-6_1

in early directives on 3D printable medical equipment FDA emergency use authorization for key equipment and designs. After an initial period of organization and process review, the US FDA has arrived at an integrated workflow for 3D printable medical equipment design review and validation. In this workflow, nonprofit organization America Makes, at the National Additive Manufacturing Innovation Institute, driven by the National Center for Defense Manufacturing and Machining, integrated the needs of the healthcare community, considered available designs from a range of sources, and matched these to the available 3D printing resources to coordinate stopgap solution response. A key part of this strategy is the development of key equipment designs by community designers, storage of these designs at the National Institutes of Health 3D Print Exchange, and review of these designs by dedicated Veterans Affairs (VA) engineers. The VA review of models for appropriateness of use in a clinical setting would result in a clinical use designation at the NIH 3D Print Exchange for the models passing VA testing. Additionally, in consultation with VA engineers, FDA would mark selected models for emergency use authorization at the NIH 3D Print Exchange. The review of designs and associated designations by VA and FDA does not guarantee that the final manufactured product would be of clinically acceptable quality, but rather makes statements on the designs and processes themselves. Of note, certain types of medical equipment, like nasopharyngeal swabs (Ford et al. 2020; Decker et al. 2020; Rybicki 2020) and surgical masks, are Class I exempt medical devices, and thus will never be formally "cleared" by the FDA. While this additional vetting is in theory an improvement over local efforts without the infrastructure, the overall efforts have varied according to expertise, experience, and comfort levels of local 3D printing community members. 3D Printing has buttressed the supply chain (Tino et al. 2020; Coté et al. 2020). This chapter reviews these experiences, with specific attention on the time period between March and May 2020 at the University of Cincinnati with team members including physicians, healthcare providers, industry, and university-based engineers working together to best deploy the available 3D printing resources in Cincinnati, Ohio, USA.

While 3D printing efforts have attempted to integrate the collective technologies in mainstream health care in North America (Rybicki 2015, 2018; Mitsouras et al. 2015, 2020; George et al. 2016, 2017; Christensen and Rybicki 2017; Giannopoulos et al. 2015; Chepelev et al. 2018; Di Prima et al. 2015), in many ways the integration is less mature than in Asia where there are longstanding relationships and trust among 3D printing groups and hospitals. Pulling groups of diverse professionals together to build devices and relationships had many success stories and challenges. In Cincinnati, the team of makers had a strong engineering representation in nearly daily meetings progressing over four months. We first detail the basic principles and general practical experiences and then focus on specific projects addressing the shortages.

2 Practical Financial Considerations

Despite the relatively high per-capita expense for healthcare in the United States (Peter 2020), dedicated resources were not available for 3D-printing-based stopgap solutions. The main consideration has been the reluctance of hospitals and local governments in paying for 3D printed medical supplies that were not regulated by the FDA. This has been a significant contributor to uncertainty around the implementation of unorthodox yet effective solutions, such as procurement of the well-publicized 3D printed conversion kits to turn snorkeling masks into N95-grade respirators. From a practical perspective, what this meant is that while there was some emergency funding to carry out engineering research, including locally in Cincinnati, there was often a disconnect between specific requests for 3D printed parts and the price per unit the requesting parties were able to reimburse. Determination of the price points itself was in question, as there were great disparities in cost among materials and the operational costs between hardware platforms, which built unreasonable expectations at times. In some cases, the design of a part required specifications that could only be achieved with certain hardware and material combinations, which translated into combinations of cost and maximal throughput that were not favorable for widespread adoption of 3D printed solutions as the mainstay of stopgap measures. For example, while the community efforts could be coordinated to create virtual 3D print farms to build face shield holders for minimal cost and at nearly sufficient throughputs, the construction of parts such as laryngoscope blades and ventilator parts required the use of limited printers capable of creating airtight or watertight models using medically compatible materials. In the end, most 3D printed stopgap solutions were financed using a combination of existing operational funds and institutional emergency procurement funds. Ultimately, the decision to proceed with the manufacturing of a specific device was thus determined on the basis of the acuity of need, the financial considerations, the limitations on technology used (e.g., biocompatible materials, specific sterilization needs), and the expected throughput accessible in the context of 3D printing.

3 Industrial Resources and Partnerships

The gestures of goodwill from the community and industrial partners played a key role in several efforts. For example, the consumer goods corporation Procter and Gamble, which has its global headquarters in Cincinnati, provided tremendous volunteerism and goodwill to our medical center during the COVID-19 pandemic. Their simple designs for personal protective equipment proved reliable and reproducible—albeit it is important to note that the tens of thousands of PPE parts donated throughout Cincinnati and beyond were ultimately not 3D printed. For larger scale manufacturing, 3D printing is not efficient for companies capable of deploying alternative established manufacturing technologies such as injection molding and other scalable technologies. The 3D printable designs could, however, provide an intermediate step in design development and testing before partners with plastic manufacturing capabilities could step in with alternative scalable manufacturing solutions. For an example of 3D printed contributions, the University of Cincinnati benefited from the local company AtriCure, which designed and 3D printed thousands of face shields that were used at the main adult teaching hospital Intensive Care Units. Other companies contributed to local efforts to fight COVID-19 by applying their engineering efforts to design simple 3D printable parts. For example, the simple 3D printable mask designed by Mark Fuller was widely considered and adopted by several enthusiasts. Consisting of a simple 3D printed frame, this mask was touted as a Do-It-Yourself project accessible to the general public—so long as the general public had access to a 3D printer. The 3D printing company Materialise developed a series of solutions, from face masks and respirators to hands-free door handles designed to minimize fomite-based transmission of COVID.

4 Community Contributions

Volunteers organized and have been tremendously helpful. For example, the community efforts in Ottawa, Canada, in conjunction with local university-based 3D printing resources produced thousands of simple face shield designs which were donated to The Ottawa Hospital and Children's Hospital of Eastern Ontario. In Cincinnati, the community volunteers countless hours for the production of essential personal protective equipment, most notably reusable masks for the patients and ear tension relief devices (O'Connor et al. in press).

Where the community-based 3D printed equipment was donated to the hospital, the local community efforts were generally organized by a small group of community coordinators, with a set hospital liaison. The hospital liaison was typically tasked with further processing of the community contributions, which typically comprised of ensuring either the final minimal stages of end-product assembly or setting up resources for material disinfection. In our experience, such disinfection typically comprised of high-level chemical disinfection guided by the chemical compatibility profiles of the materials in question.

5 Device Sterilization and Reuse

The practical aspects of device sterilization must be addressed at the initial stages of device design considerations, as the required level of disinfection or sterilization will guide material selection and factor into throughput and costs. Briefly, there are three general levels of removal of unwanted bacterial and chemical contamination: cleaning, disinfection, and sterilization. Cleaning is typically the first step; it refers to the removal of macroscopic foreign material from the sterilized part, typically using enzymatic agents, solvents, or detergents. Cleaning is typically followed either by disinfection or sterilization. In disinfection, either most or all living pathogens are removed from the surface of the disinfected material, depending on whether disinfection is low-level or high-level, respectively. In chemical disinfection, the degree of disinfection is controlled by the concentration of the disinfectant and the exposure time. Ultimately, some forms of chemical disinfection, after sufficient exposure and at sufficiently high concentrations of the agent used, can be termed 'chemical sterilization'. In most applications with skin and mucosal surface contact, high-level disinfection is sufficient. Where contact with sterile tissues is anticipated, sterilization of the involved device is indicated. Chemical sterilization uses agents such as concentrated hydrogen peroxide for a specific amount of time to destroy all surface pathogens. More commonly used forms of sterilization are based on physical techniques—specifically, the exposure of sterilized parts to steam at up to 132 °C for 40 min is typically sufficient to achieve sterilization of most items. In the context of some printing materials, such as thermoplastic polyurethane (TPU), exposure to such high temperatures is undesirable as it may result in dramatic changes in the properties and the configuration of the 3D printed parts. For this reason, chemical sterilization or chemical high-level disinfection may be preferred, using guidance derived from a combination of manufacturer-supplied chemical compatibility charts for key components of the stopgap 3D-printed part and the chemical agent-specific instructions for achieving the desired degree of disinfection or sterilization.

6 Specific Implementation Examples

6.1 Personal Protective Equipment: Face Masks

Early 3D printing during the pandemic focused on face masks and filter carriers. Community designs based on the properties of thermoplastic polyurethane (TPU) were among the earliest designs identified, modified, and tested. The attractive features included the putative ability to conform the masks to the wearer's face in order to ensure the best fit, and the ability to print these masks without significant expenditure of support material since the ability to bend these masks after printing meant that the printed flat mask could be folded after heating into its final configuration. Finally, these masks could be created using desktop printer farms or outsourced to community volunteers. We experienced several practical challenges with TPU-based masks. Most importantly, the prints available to us were not reliable, in that following assembly, the masks were not airtight; this limited utility significantly. The flexibility of the material, especially following heat exposure, limited its utility in practical scenarios. Limited material procurement was also a barrier, and ultimately no TPU-based foldable designs were delivered to our front-line healthcare workers. We considered the manufacture and adaptation of full-face snorkeling mask conversion kits. While technically promising and eventually passing N95 testing, such solutions were complicated by the limited availability of the funds allocated to the purchase of snorkeling masks and the limited availability of stock of such masks at the scales necessary to fill the medical demand. We also have problems obtaining adaptable filter and filter material. We, therefore, considered the application of a wide range of designs. For instance, we prototyped individualized designs based on face scanning with conformal dome-shaped masks with replaceable filter carriers at the anterior aspect of the mask. These designs had significant associated material and printing time costs. We considered hybrid masks with minimal 3D printed frames where the seal was formed by crafting rubber tubing or door/window seal material around the expected location of the facial seal to ensure an airtight fit. The greatest challenge with manufacturing these masks at sufficiently large scales has been the selection of a filtration material capable of providing sufficient protection while ensuring the comfort of the wearer. For these purposes, materials ranging from combinations of fabrics, paper towels, surgical cover materials that are typically discarded, and even vacuum cleaner bags were investigated for use in viral filtration. We applied local expertise at the University of Cincinnati to quantitatively examine the filtration potential of such materials and were able to objectively identify the optimal combinations of materials for our mask designs from the standpoint of availability and comfort. Ultimately, several hundred masks of different types were produced as manufacturing by the industry ramped up (Fig. 1).

Of note, parallel community efforts in Cincinnati involved the development and production of masks for the wider general population, by various community groups organizing for a response. These groups benefitted from 3D printable tools to facilitate mask production, but generally functioned autonomously and were ultimately capable of distributing thousands of simple cloth masks to the general population.

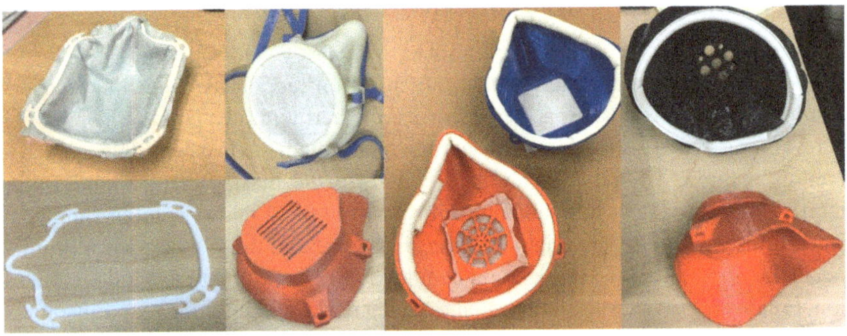

Fig. 1 Various mask designs evaluated or refined at our laboratories, ranging from minimal 3D printed frame-based ones to fully 3D printed filter carriers

6.2 Personal Protective Equipment: Face Shields

Various 3D printing designs were evaluated in the context of a rapid 3D printing response. However, given the tremendous numbers of face shields required, the requirement for reusability, and the limited throughput potential given the existing resources, injection molding-based manufacturing was preferred. While community efforts did significantly contribute to providing face shields through coordination by community organizers, creating a virtual desktop printer farm, such efforts often required separate procurement of pre-cut transparent polycarbonate shields to couple with the 3D printed holder. Given a significant and unexpected surge in demand, the strain placed on commercial entities capable of providing such materials often resulted in delays for Canadians, and to a certain extent in the United States as well.

The state of Ohio has arranged to acquire and distribute at least one million reusable face shields in the early stages of the pandemic. These shields were developed in collaboration with several Ohio organizations. The State of Ohio committed to maintaining this stockpile and to coordinating its distribution through the MAGNET Ohio Manufacturing Alliance and the Ohio Hospital Association. The design ultimately supported was identical to that being produced by Proctor & Gamble and consisted of three parts, two of which were injection molded. The parts underwent engineering evaluation with extensive testing, and it was confirmed that these could stand many use cycles. The contribution of commercial partners has been indispensable in reaching this goal.

6.3 Nasopharyngeal Swabs

While limited throughput capabilities for 3D printing of nasopharyngeal swabs were available at the University of Cincinnati, it was clear that broader commercial partnerships were necessary to ensure sufficient supplies for the wider needs of the city. There are many swab designs with a focus on the design initiated at the University of South Florida; some of the controversies surrounding swab printing have been discussed (Rybicki 2020). The 3D printed swabs have been demonstrated to be either equal to, or to outperform existing commercially available swabs and were capable of obtaining enough viral particles to ensure a confident diagnosis of COVID-19 infection.

6.4 Environmental Modifiers

Simple modifiers in behavior or equipment can often yield surprisingly positive results. A shining example of this principle has been the institution of handwashing, proposed in modern medicine initially by Semmelweis, and likely responsible for millions of lives saved since widespread adoption. Simple 3D printed devices have emerged to capitalize on this principle. Our groups have been involved in the manufacturing and distribution of several such devices, and we will use door handle openers and ear savers as examples of these efforts.

Inanimate objects capable of transmitting pathogens are referred to as fomites. In hospital environments, there are dozens of door handles that patients and clinicians touch on the way to their appointments and duties. While at destination, hand washing is tremendously helpful in reducing fomite transmission. Unfortunately, unconscious reflexes may result in hand–face contact and possible pathogen transmission on the road to the destination. To help address this, multiple groups, most notably engineers at Materialise (Leuven, Belgium), have proposed 3D printable door openers. At the University of Cincinnati, we were able to rapidly capitalize on such designs to manufacture hundreds of door openers and distribute these two key areas within the hospital (Fig. 2).

Similarly, wearing a masks is very strongly recommended to limit the spread of COVID-19. In our experience, the incessant wearing of a mask can be a tremendous source of discomfort, focused on posterior ears, and resulting in some personnel removing the mask for extended periods of time. Simple designs of the so-called "ear-savers" are easy and cheap to manufacture in bulk and often provide significant relief to posterior ear pain. We have distributed thousands of such devices broadly, with only minimal time and capital investment for production.

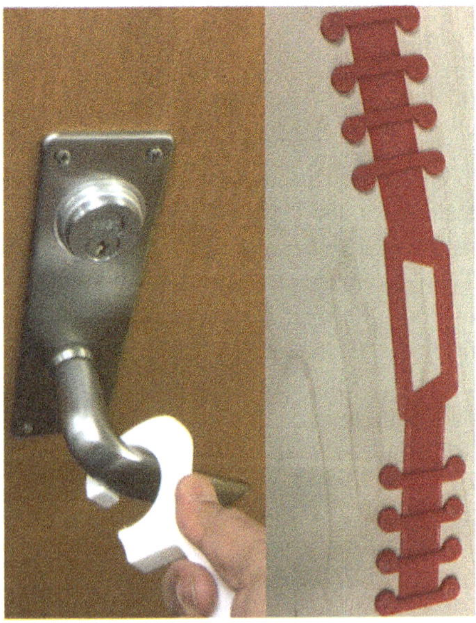

Fig. 2 Examples of a door handle opener (left) and ear saver (right)

6.5 Specialized Equipment: Laryngoscope Blades

A surprisingly unforeseen shortage was encountered when reports of limited stocks of video-assisted laryngoscopy equipment emerged. Laryngoscopes are devices used for the intubation of patients. Specifically, laryngoscopes help clear the pathway for the placement of a tube used to deliver oxygen to the lungs (via an endotracheal tube). If endotracheal tube placement is conducted blindly, the risks include damage of the laryngeal structures and intubation of the stomach with the resultant aspiration of gastric contents by the patient, partial airway compromise, and transport of gastric contents onto the intubating physician. While many patients with typical anatomic configuration of the mouth and neck can be intubated using direct visualization with a reusable metallic laryngoscope, some patients have neck anatomy that precludes direct visualization. For these cases, video laryngoscopy has been devised. In video laryngoscopy, a curved reusable blade contains a hollow cavity where a reusable camera is fixed in place (Figs. 3 and 4). Laryngoscopy blades typically cannot be reused.

The design of these blades is subject to significant constraints. First, the material should be biocompatible. Second, the part must be watertight. Third, there must be a completely transparent window that protects the sensitive camera from vapors and secretions to ensure optimal visualization and prolong the camera lifecycle. Fourth, during the limited supply chain during the pandemic, the blade would ideally be reusable. At our laboratories, we developed stopgap blades by using PolyJet printing with medically compatible materials. Adapting the initial overall design from segmentation of a CT scan of the existing laryngoscope blade, we carried out numerous design and testing iterations on placing a window at the end of the laryngoscope, in a 6 × 6 mm footprint, with the inability to polish or buff the surface to ensure transparency. Our final design included the placement of a polycarbonate sheet into the window, held in place by traps within a specially designed canal.

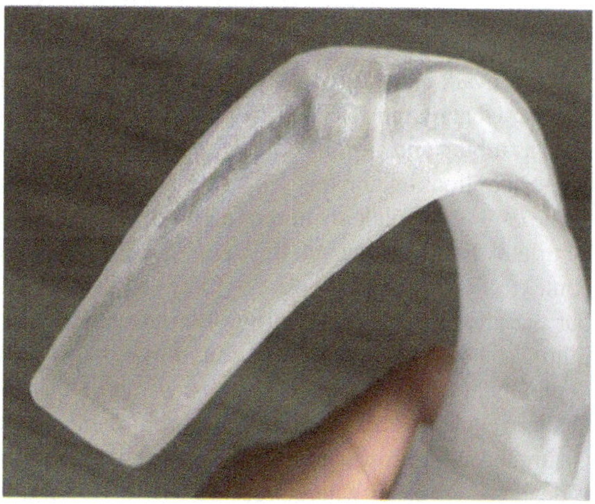

Fig. 3 The tip of a 3D printed laryngoscope blade with the expected location of the camera window. Note the lack of window transparency

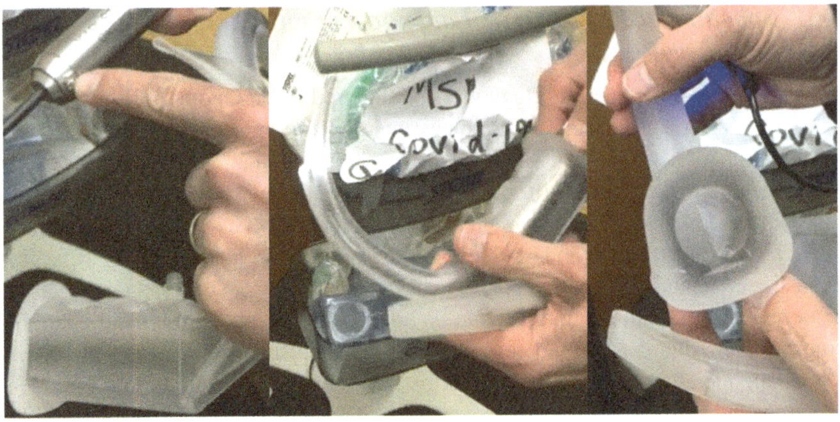

Fig. 4 The fitting of video laryngoscopy camera within the 3D printed laryngoscope prototype. A notch on the camera (left) closely fits into a groove on the blade (right), ensuring an adequate tight fit of the camera within the blade (center)

Extensive testing confirmed reusability under chemical disinfection conditions and the complete absence of mechanical motion of the entrapped polycarbonate window. Operating with batches of 12 laryngoscope blades, we were able to deliver up to 40 blades per week. Fortunately, last-minute procurement deals staved off the need for 3D printed blades in favor of a complete replacement of the video laryngoscopy platform at the hospital level.

6.6 Specialized Equipment: Ventilators and Ventilator Parts

Our teams prototyped several ventilator designs. The designs evaluated were based initially on replicating the general principles employed by established commercial ventilators such as Go2Vent (Vortran, Sacramento, USA) and later by employing publicly available designs distributed by the Illinois Rapid Vent (University of Illinois Urbana-Champaign 2020). While the development and 3D printing of simpler parts such as tubing and various connectors for the use of existing equipment with a wider range of options of adapters, filters, and suction devices have significantly contributed to the care of the patients in the Intensive Care Units, a fully functional 3D printable ventilator likely represents the apex of the medical equipment design space in the context of COVID. Not only are there requirements for the material to be airtight, but the allowable manufacturing tolerances are quite small, with the potential for injury or loss of life being quite high in either the inappropriate manufacturing or deployment of experimental designs. To provide an illustrative example, one of the challenges addressed by our team during the design refinement phase has been to ensure that a high-pressure pop-off valve functions appropriately. This pop-off valve is designed for cases where pressure within the ventilator exceeds 60 mm Hg, which can result in barotrauma and severe lung injury. Using the provided schematics and our manufacturing technology, the 3D printed pop-off valve was initially too loose, falling off the ventilator assembly spontaneously, and following refinement it was seated too tightly, with pressures necessary to pop this valve likely exceeding 60 mm Hg. While we manufactured and prototyped a working 3D printed ventilator, appreciating the complexity of the task and the limited demand in our institutions, our groups did not pursue completely replacing ventilators, until such a time that doing this would be unavoidable. With the ramping up of manufacturing that happened since the outbreak of the pandemic, we are hopeful that the time for locally 3D printed ventilators will never come.

6.7 Facilitating Coming Back to School

While much of the literature and experiences cited fell in the spring and summer 2020, with the arrival of September and the Fall Season in North America, there is a traditional return to school. This comes with new responsibilities during COVID-19, and new challenges to design parts such as face masks that are durable and comfortable. Face coverings are usually mandatory on school campuses, but those policies are difficult to enforce. Creating and validating accessible 3D printable Do-It-Yourself (DIY) masks and encouraging people in the community to supply those remains important, especially for individuals with difficulties in tolerating widely available designs or in areas where a local shortage arises. At the University of Cincinnati, while our leadership appreciated the presented mask designs, only limited prototypes were procured for further evaluation. The reluctance for a wider adoption may be in part attributable to a lack of an established and widely accepted reusable low-cost standard mask. The role of 3D printing in this space may be limited in the future. First, the ramping up of manufacturing of traditional designs will hopefully belay the need for such a standard, and if a standard versatile 3D printable design emerges and is widely adopted, we expect the deployment of large-scale manufacturing technologies such as injection molding-based manufacturing.

7 Conclusion

The versatility of 3D printing has been tremendously useful in addressing some of the key equipment shortages with reasonable stopgap alternatives. Interdisciplinary teams included hospitals, universities, engineers, community, and industry partnerships. Careful consideration of the design characteristics, the available throughput, and the acuity of need in consultation of stakeholders helped to maximize the impact of 3D printing and shift the burden of projects requiring high throughput to alternative technologies. Simply engineered PPE and other products proved essential for supporting the frontline during the COVID-19 pandemic and rapidly addressing deficiencies on the way to the ongoing recovery. Engineers require integration into tight working groups with healthcare providers and hospital administrators to have the highest impact. Without those connections, product designs run the risk of remaining in isolation. We hope we have provided compelling examples of how our teams found synergies among professionals in engineering (both from academia and industry), medicine, and the larger community to help limit coronavirus spread. In all cases, the practical use of 3D printing has been that of temporizing the acute equipment needs until traditional manufacturing has been able to ramp up. We sincerely hope that 3D printing is never again necessary to fill this role in the setting of widespread equipment shortages and the tragic loss of life.

References

Chepelev L, Wake N, Ryan J, Althobaity W, Gupta A, Arribas E, Santiago L, Ballard DH, Wang KC, Weadock W, Ionita CN, Mitsouras D, Morris J, Matsumoto J, Christensen A, Liacouras P, Rybicki FJ, Sheikh A (2018) RSNA special interest group for 3D printing. radiological society of North America (RSNA) 3D printing special interest group (SIG): guidelines for medical 3D printing and appropriateness for clinical scenarios. 3D Print Med 4:11

Christensen A, Rybicki FJ (2017) Maintaining safety and efficacy for 3D printing in medicine. 3D Print Med 3(1):1

Coté JJ, Haggstrom J, Vivekanandan R et al (2020) COVID-19 and a novel initiative to improve safety by 3D printing personal protective equipment parts from computed tomography. 3D Print Med 6:20. https://doi.org/10.1186/s41205-020-00073-6

Decker SJ, Goldstein TA, Ford JM et al (2020) 3D printed alternative to the standard synthetic flocked nasopharyngeal swabs used for COVID-19 testing. Clin Infect Dis ciaa1366. https://doi.org/10.1093/cid/ciaa1366. Epub ahead of print. PMID: 32910817; PMCID: PMC7499529

Di Prima M, Coburn J, Hwang D, Kelly J, Khairuzzaman A, Ricles L (2015) Additively manufactured medical products—the FDA perspective. 3D Print Med 2(1):1–6

Ford J, Goldstein T, Trahan S et al (2020) A 3D-printed nasopharyngeal swab for COVID-19 diagnostic testing. 3D Print Med 6:21. https://doi.org/10.1186/s41205-020-00076-3

George E, Liacouras P, Rybicki FJ, Mitsouras D (2017) Measuring and establishing the accuracy and reproducibility of 3D printed medical models. Radiographics 37(5):1424–50583

Giannopoulos AA, Chepelev L, Sheikh A, Wang A, Dang W, Akyuz E, Hong C, Wake N, Pietila P, Dydynski PB, Mitsouras D, Rybicki FJ (2015) 3D printed ventricular septal defect patch: a primer for the 2015 radiological society of North America (RSNA) hands-on course in 3D printing. 3D Print Med 1:3

Giannopoulos AA, Mitsouras D, Yoo SJ, Liu PP, Chatzizisis YS, Rybicki FJ (2016) Applications of 3D printing in cardiovascular diseases. Nat Rev Cardiol 13(12):701–718. https://doi.org/10.1038/nrcardio.2016.170

Mitsouras D, Liacouras P, Imanzadeh A, Giannopoulos AA, Cai T, Kumamaru KK, George E, Wake N, Caterson EJ, Pomahac B, Ho VB, Grant GT, Rybicki FJ (2015) Medical 3D printing for the radiologist. Radiographics 35:1965–1988

Mitsouras D, Liacouras PC, Wake N, Rybicki FJ (2020) RadioGraphics update: medical 3D printing for the radiologist. Radiographics. 40(4):E21–E23. https://doi.org/10.1148/rg.2020190217

O'Connor Z, Huellewig D, Sithiyopasakul P, Morris JA, Gan C, Ballard DH (in press) 3D printed mask extenders as a supplement to isolation masks to relieve posterior auricular discomfort: an innovative 3D printing response to the COVID-19 pandemic 3D Print Med

Peter G (2020) Peterson foundation. The United States is an Outlier in Healthcare Spending. https://www.pgpf.org/blog/2020/07/how-does-the-us-healthcare-system-compare-to-other-countries

Rybicki FJ (2015) 3D printing in medicine: an introductory message from the Editor-in-chief. 3D Print Med 1:1. https://doi.org/10.1186/s41205-015-0001-5

Rybicki FJ (2018) Medical 3D-printing and the physician-artist. Lancet 391(10121):651–652. https://doi.org/10.1016/s0140-6736(18)30212-5

Rybicki FJ (2020) 3D printing in medicine: COVID-19 testing with 3D printed nasopharyngeal swabs. Clin Infect Dis. https://doi.org/10.1093/cid/ciaa1437. Epub ahead of print. PMID: 32949233

Tino R, Moore R, Antoline S et al (2020) COVID-19 and the role of 3D printing in medicine. 3D Print Med 6:11. https://doi.org/10.1186/s41205-020-00064-7

University of Illinois Urbana-Champaign (2020) Illinois RapidVent. https://rapidvent.grainger.illinois.edu/

The Role of Distributed Manufacturing and 3D Printing in Development of Personal Protective Equipment Against COVID-19

John Scott Frazer

Abstract Global shortages of personal protective equipment (PPE) have been at the forefront of public attention throughout the recent crisis. Initial lack of information regarding the route of transmission of the virus fuelled panic buying of face masks and gloves, as well as shortages in healthcare facilities. Home-made cloth masks have become commonplace, and makers have seized upon opportunities to increase capacity by producing PPE with 3D printing. Mask designs typically incorporate flexible filament face pieces with an integrated filter, but devices to improve the fit of simple surgical face masks have also been produced, and are not burdened with such strict testing requirements. Manufacture of face shields has been a triumph for the 3D printing community, with the availability of highly optimised and tested designs which are printable on most domestic machines, comfortable to use, and keep production time to a minimum. Across the globe, these devices have been manufactured locally and delivered to nearby hospitals by volunteers. Thus, the COVID-19 pandemic has spurred development in the field of distributed manufacturing, the long-awaited holy grail of home 3D printing. In the following chapter, we provide an overview of currently available 3D printed PPE, reviewing the designs for durability, ease of use, and clinical effectiveness, as well as exploring future directions for PPE manufacture by 3D printing.

Abbreviations

FDM	Fused Deposition Modelling
SLA	Stereolithography
SLS	Selective Laser Sintering
MJF	Multi-Jet Fusion
PPE	Personal Protective Equipment
PAPR	Powered Air-Purifying Respirator

J. S. Frazer (✉)
Somerville College, University of Oxford, Oxford, UK
e-mail: drjsfrazer@gmail.com

K. Sandhu et al. (eds.), *Emerging Applications of 3D Printing During CoVID 19 Pandemic*, Lecture Notes in Bioengineering, https://doi.org/10.1007/978-981-33-6703-6_2

ABS	Acrylonitrile Butadiene Styrene
PLA	Polylactic Acid
PEEK	Polyether Ether-Ketone
RT-PCR	Reverse-Transcriptase Polymerase Chain Reaction

1 Introduction

COVID-19 refers to the clinical syndrome caused by the SARS-CoV2 virus, a respiratory virus that invades a host through the nasal mucosa. From there it goes on to cause primarily respiratory symptoms, although many other systems may be involved in the critically ill (Siripanthong et al. 2020; Ronco et al. 2020; Frazer and Everden 2020). Initial guidance suggested that the primary mechanism of transmission was via droplet spread. However, more recent evidence points towards multiple other routes of transmission, including aerosol and ocular spread. During a respiratory pandemic such as this, healthcare workers are at particular risk; approximately 20% of victims of severe acute respiratory syndrome (SARS) were healthcare workers, and up to 29% of COVID-19 transmissions in hospital involve this group (Wang et al. 2020; World Health Organisation 2020c. While a recent systematic review has demonstrated that social distancing, eye protection, and face masks are all effective in reducing transmission of COVID-19 (Chu et al. 2020), lack of knowledge and inconsistent use of personal protective equipment (PPE) was cited as a contributing factor in an outbreak of SARS amongst healthcare workers in one case series in Toronto, Canada (Ofner-Agostini et al. 2006). These startling facts highlight the urgent need for effective and easily available PPE during the COVID-19 pandemic (Frazer et al. 2020).

Given the unprecedented number of cases worldwide, stocks of PPE have been rapidly depleted. Prior to the COVID-19 pandemic, China supplied half of the world's face masks, and was the sole mass producer of personal protective gowns (Burki 2020). Burki cites the alignment of Chinese New Year, which often produces a predictable hiatus in production, with their initial lockdown and subsequent embargo on international export of PPE as a key exacerbating factor in the global shortage (Burki 2020). WHO has warned that 89 million medical masks, 76 million examination gloves, and 1.6 million pairs of goggles will be required globally each month in order to meet the peak of COVID-19 safely (World Health Organisation 2020b). This increased demand has resulted in the price of surgical masks increasing by a factor of 6, N95 masks increasing by a factor of 3, and surgical gowns by a factor of 2. While China has now dramatically increased the production of surgical masks from 20 million per day before the pandemic to 110 million per day, intense strain on supply chains remains, and any future Chinese lockdown may further restrict the supply of critical items internationally. Many countries including the UK and USA have previously considered having a minimal stockpile of critical medical supplies as a sound strategy, although with the tight restrictions placed on PPE export by individual countries, local manufacture and stockpiling of these valuable devices is now becoming a clearly attractive option (Feinmann 2020). Thus, it is critical to develop distributed manufacturing strategies which will enable de-centralised production of PPE in individual countries, and even domestically. Indeed the model of distributed manufacturing is powerful in the face of a pandemic, and is ideally suited to the production of face shields and other simple devices. This approach has the potential to reduce waste and emissions, as stockpiles of resources are less necessary, and transport of completed parts over long distances is not required (Tarfaoui et al. 2020).

The open-source and 3D printing community has responded to the urgent need for PPE with a number of innovative solutions, including some devices currently in widespread use in hospitals and healthcare facilities throughout the world. This chapter will begin with a summary of the possible mechanisms of transmission of SARS-CoV2, and the various classes of PPE used to guard against this, going on to discuss the various innovative 3D printed solutions to the PPE crisis, and potential future directions of the work.

2 SARS-CoV2 Transmission and Infection

The first study reporting human-to-human transmission of SARS-CoV2 was published on 24th January 2020 (Chan et al. 2020). Since then, more than 50 million individuals have been infected (Johns Hopkins University of Medicine Coronavirus Resource Centre 2020). Much of the early epidemiological data regarding transmission was based on SARS-CoV, the causative agent of SARS. Transmission was originally thought to occur predominantly via respiratory droplets, with aerosolisation of viral particles only occurring during 'aerosol generating procedures' (AGPs), such as endotracheal intubation and bronchoscopy (Peiris et al. 2003). This mechanism of transmission is under constant revision and debate, however (Jayaweera et al. 2020). Even the accepted list of AGPs is not clear: a 2012 systematic review found that endotracheal intubation, non-invasive ventilation, tracheostomy, and manual ventilation before intubation were associated with a significantly higher risk of transmission of respiratory infections, while 'intubation-associated procedures', endotracheal aspiration, sputum collection, suctioning, bronchoscopy, nebuliser treatment, administration of standard or high-flow oxygen and manipulation of masks, defibrillation, chest compression, and nasogastric tube insertion were not significantly associated with increased risk of infection (Tran et al. 2012). Endotracheal intubation is consistently rated as one of the highest risk AGPs (Cook 2020). However, some authors even surmise that supplemental oxygen above a flow rate of 6 L/min should be considered an AGP (Cheung et al. 2020).

2.1 Droplet Transmission

SARS-CoV2 is thought to be mainly transmitted by respiratory droplets. Infected individuals with high viral load in their nasopharynx and upper airways cough or sneeze, producing droplets that travel short distances, and either land directly on the mucosa of nearby individuals, or on surfaces. Individuals who touch these surfaces may then self-inoculate by touching their mouth or nose. Viable viral particles have been detected on plastic and stainless steel even after 72 hours (half-life of 5.6 and 6.8 hours respectively), although surfaces such as copper appear to fare better, with no viable particles detected after approximately 4 hours (van Doremalen et al. 2020).

2.2 Aerosol Transmission

The relatively low replication factor of SARS was initially thought to indicate that its spread is primarily via droplets landing on those in close proximity to an infected person, rather than via aerosolisation (World Health Organisation 2020a. However, subsequent analysis of mass outbreaks determined that aerosol spread was indeed likely (Yu et al. 2004). The definition of an aerosol varies, but commonly includes reference to particles small enough to remain in the air for extended periods; ranges from 5 to 20 μm in diameter have been proposed (Nicas et al. 2005; Siegel et al. 2007). Aerosol particles are said to linger in the air longer than droplets, and the smaller aerosol particles are able to travel deeper into the smaller airways of the lungs where they lead to infection (Thomas 2013). SARS-CoV and SARS-CoV2 viral particles aerosolised under experimental conditions by nebulisation remained viable in the air even after 3 hours, with only a small reduction in viral titre by the end of this period (van Doremalen et al. 2020). A half-life of 1.1–1.2 hours for viral aerosols was calculated. That people who are in a confined space for a long period, or share a living space are more likely to be infected favours droplet spread. However, there are now multiple reports of infections in which the most likely mode of transmission was aerosolisation: in March 2020, a USA-based choir was the site of a major aerosol-based transmission (Hamner et al. 2020). Sneezes are thought to be aerosol-generating (Cheung et al. 2020), and LASER light scattering experiments have revealed that even normal conversation is capable of generating particles small enough to be aerosolised (Stadnytskyi et al. 2020). A pre-print article claims to have detected SARS-CoV2 RNA simply in the exhaled breath of patients suffering from COVID-19 (Ma et al. 2020) and there is even some evidence of detection of aerosolised RNA in hospitals in China (Liu et al. 2020), although to date no experiments have proven that this is indicative of viable viral particles. In response to the risk of aerosolisation, some authors suggest that in addition to wearing face coverings in public and scrupulous hand washing, overcrowding should be avoided, adequate ventilation provided in indoor spaces, and further airborne infection control procedures such as air filtration and ultraviolet exposure implemented (Morawska and Milton 2020; Morawska and Cao 2020). Some suggest that these factors may be even more important than individual PPE (Morawska et al. 2020).

2.3 *Ocular Transmission*

Some recent work is beginning to highlight the eye as an important route of viral shedding and entry. Other viruses in the coronavirus family are known to transmit and infect via the eyes (Seah and Agrawal 2020). There are even some reports of possible COVID-19 infection via this route (Lu et al. 2020). One case series from Hubei province, China, reported 31.6% of their 38 patients with confirmed COVID-19 had ocular manifestations of the illness, including conjunctivits, increased secretions, and others (Wu et al. 2020). They also reported that 5.1% of their subjects had detectable SARS-CoV2 RNA in their eye secretions. This is possibly as a result of the nasolacrimal system facilitating transfer of viral particles from the nasopharynx to the eye, although debate is ongoing. A prospective study analysed a small sample of patients with COVID-19 without eye symptoms, finding both RT-PCR and viral culture of eye secretions negative for SARS-CoV2 (Seah et al. 2020). A retrospective analysis of patients hospitalised in Suizhou, China, found that the proportion of patients with COVID-19 who wore glasses for more than 8 hours per day (5.8%) was much lower than the population as a whole (31.5%) (Zeng et al. 2020), possibly hinting at a protective effect of wearing glasses against transmission, although further work is required to clarify this.

3 3D Printing of Personal Protective Equipment

An impressive and broad array of designs for 3D printed PPE are available online. Multiple creative solutions exist, including devices to hold the cloth over a user's face to act as a filter (ctwiles 2020), and even devices for hands-free opening of doors (François et al. 2020). A 3D printed surgical mask ear protector has been designed by a USA-based Boy Scout, and has now become widespread (Interesting Engineering 2020). An innovative solution to self-isolation and hospital crowding has been proposed by manufacturers in China, who have created a 10 square metre cabin which can be 3D printed using a bespoke construction-grade printer loaded with construction waste products and sand (Global Construction Review 2020). There have even been some attempts to 3D print gloves for use during swimming using flexible filament, although to date this has not been attempted as a route to manufacture PPE (RCLifeOn 2017). The most common 3D printed PPE devices include face shields, face masks, and parts for powered air-purifying respirators (PAPRs). These will be discussed in detail below and mentioned in Table 1.

Table 1 List of characteristics of PPE described in this chapter

Class of PPE	Device	Printing method	Material	Comments	References
Face shields	3D4care face shield	FDM	ABS/PLA + PVC transparent film	Validated for use by surveying interventional radiologists	Sapoval et al. (2020, 3D4Care (2020)
	Prusa RC1	FDM	ABS/PLA + PVC transparent film		Wesemann et al. (2020), Prusa Research 2020), de Araujo Gomes et al. (2020), Lemarteleur et al. (2020)
	Prusa RC2	FDM	ABS/PLA + PVC transparent film	Stackable print design facilitating rapid production	Wesemann et al. (2020), Prusa Research (2020), de Araujo Gomes et al. (2020), Lemarteleur et al. (2020)
	3D Face Shield V3 (Budmen Industries)	FDM	ABS/PLA + PVC transparent film		Wesemann et al. (2020)
	Easy 3D Face Shield (Honach Hermann)	FDM	ABS/PLA + PVC transparent film	Rated highest overall in terms of user protection by Wesemann et al.	Wesemann et al. (2020)
	Prusa PRO	FDM	ABS/PLA + PVC transparent film	Reportedly has achieved compliance with European PPE regulations	Prusa Research (2020), Armijo et al. (2020)

(continued)

Table 1 (continued)

Class of PPE	Device	Printing method	Material	Comments	References
	Neijhoft design	FDM	ABS/PLA + PVC transparent film	Skeletonised version of Prusa shield, intended to improve print speed and reduce plastic use. When printed in PLA, it can be heated with warm water for further manipulation of frame	Neijhoft et al. (2020)
	Chaturvedi design	FDM	ABS/PLA + PVC transparent film	PVC film is clipped to frame via metal clips in order to reduce time spent punching holes in the film	Chaturvedi et al. (2020)
	Viera-Artiles design	FDM	ABS/PLA + PVC transparent film	Includes head light holder	Viera-Artiles and Valdiande (2020)
	Maracaja design	SLA	SLA resin 'Tough 1500' and 'Draft Resin' used in study	Transparent shield attached to the hollow tube to facilitate airflow over wearer's face	Maracaja et al. (2020)
Face masks	CEG Extreme	FDM	ABS/PLA	Frame for holding filtration material close to face	ctwiles (2020)
	Swennen design	SLS	Polyamide composite PA11-SX 1450	Utilised 3D scanning to provide precise fit of mask to user's face. Print time 12 hours. No formal fit testing	Swennen et al. (2020)

(continued)

Table 1 (continued)

Class of PPE	Device	Printing method	Material	Comments	References
	Copper3D NanoHack mask	FDM	ABS		#HackThePandemic (2020)
	Pan Fab	FDM	ABS/PLA	Adjunct to existing N95 mask reporting an increased chance of success of fit testing. Requires metal wires in addition to 3D printed parts	McAvoy et al. (2020)
	Wiles COVID pandemic mask	FDM	PLA plus proprietary filter material (NanoNet and NanoForce by Cummins)	Fifty masks were customised with wearer's names, and used in a hospital in the United States of America to good effect	Thomas et al. (2020), ChrisWilesDO (2020)
	Stopgap surgical face mask (SFM) revision B	MJF/SLS/SLA	Powder-bed nylon (MJF/SLS), or BioMed Clear (SLA)	Specialist printer and materials required	VHA Innovation (2020)
	Imbrie-moore design	SLA	Biocompatible silicone and multipurpose polyurethane	Requires an existing N95 mask to be cut into quarters, with the 3D printed parts utilising the quartered N95 material as a filtration medium	Imbrie-Moore et al. (2020)

(continued)

Table 1 (continued)

Class of PPE	Device	Printing method	Material	Comments	References
	Maker Mask	FDM	ABS/PLE/PEKK	Printed by Skrzypczak et al. using a specially designed high-temperature RepRap-class 3D printer capable of printing PEKK plastic	Maker Mask and Respitator (2020), Skrzypczak et al. (2020)
	Direct printing of N95 mask materials	Specialist methods	Polypropylene and styrene-(ethylene-butylene)-styrene	Still in experimental stages. See referenced review article	Ishack and Lipner (2020)
Mask seal	Custom face seal for N95 mask	FDM	ABS	A model of the user's face is built by 3D scanning, and the seal then custom printed	Cai et al. (2018)
Powered air-purifying respirators	Erickson manifold	SLA	Formlabs 'durable' resin	Manifold used to retrofit filters to disused arthroplasty helmets to form a PAPR	Erickson et al. (2020)
	Hubbard design	FDM	ABS/PLA	Powered filter holder and fittings used to retrofit firefighter hood to form a PAPR	Hubbard and Pearce (2020)
	CT scanned parts	FDM	ABS/PLA	Existing parts for PAPRs were scanned using a CT scanner, and parts replicated by 3D printing with impressive accuracy	Coté et al. (2020)

(continued)

Table 1 (continued)

Class of PPE	Device	Printing method	Material	Comments	References
Snorkel mask adapters	EasyBreath® adapter	FDM	ABS/PLA	Device to allow fitting of filter to EasyBreath® snorkel mask	Thierry et al. (2020), Germonpre et al. (2020)
	Decathlon adapter	SLA	Formlabs Draft Resin	Device to allow fitting of filter to Decathlon snorkel mask	Formlabs (2020)
	Ocean Reef ARIA Protection Adapter	SLA	SLA resin	Device to allow fitting of filter to ocean reef ARIA mask	Germonpre et al. (2020)
Isolation cabin	Isolation cabin	Bespoke concrete 3D printer	Sand plus construction waste	Several have been delivered to hospitals in China	Global Construction Review (2020)
Door opener	Hands-free door opener	FDM	ABS/PLA		François et al. (2020)
Ear protector	Surgical mask ear protector	FDM	ABS/PLA		Interesting Engineering (2020)
Gloves	Webbed swimming gloves	FDM	Flexible filament	Developed as a swimming adjunct, not currently tested or in use as PPE	RCLifeOn (2017)

3.1 3D Printed Face Shields

Face shields are an attractive target for additive manufacture and 3D printing. They are often simple devices, and their effectiveness is obvious: any impermeable and transparent shield held in front of the eyes and mouth will be functional. Indeed, face shields have been shown to reduce the risk of droplet and aerosol contamination of the wearer's face (Lindsley et al. 2014), although it should be noted that they are not considered by themselves sufficient PPE in the absence of an appropriate face mask (Roberge 2016). As the devices are not directly in contact with patients, they are also free from many of the regulatory burdens present in other areas. Severe shortages in the USA sparked the relaxation of regulations surrounding PPE (and face shields in particular), with the FDA stating that they did not object to distributed manufacture of these devices during the pandemic (Flanagan and Ballard 2020). It is noteworthy that shields may have a direct effect on supplies of other PPE; reduced soiling of masks protected by the shields increases their lifespan, possibly reducing overall numbers used (Centres for Disease Control and Prevention 2019).

3D printed face shields are one of the major success stories of distributed manufacturing during the pandemic, with hundreds of the devices reportedly produced daily (Griffiths et al. 2020), and feedback from front-line healthcare workers facilitating modification of the devices in almost real time. A device designed by 3D4care in collaboration with the University Hospitals of Paris to protect interventional radiologists during procedures was rated very highly amongst wearers (Sapoval et al. 2020), and has subsequently been deployed to more than 300 healthcare facilities via a consortium of companies who banded together to produce them (3D4Care 2020). The headband and 3D printed front piece is reusable, thus minimising plastic waste, while the transparent plastic sheet is disposable. The device may be disinfected by immersion in 0.5% sodium hypochlorite solution for 15 minutes. Four popular face shield devices, the Prusa RC1 and RC2, as well as the 3D Face Shield V3 (Budmen Industries), and the Easy 3D Face Shield (Honach Hermann) were compared for practical use on an intensive care unit and in a dental office (Wesemann et al. 2020). The study concludes that 3D printed shields are valuable in protecting healthcare workers against exposure to pathogens, and the authors make several recommendations regarding the requirements of a face shield design based on user feedback; the shield must permit wearing of other PPE underneath (such as a face mask or goggles), and must cover at least to the wearer's ears. The authors finally recommend the Easy 3D Face Shield overall in terms of user protection, although they make the important point that scalability of this device is difficult due to its non-stacking design, unlike other devices such as the Prusa RC2.

The Prusa RC2 and RC3 devices are among the most commonly printed face shields (Prusa Research 2020). The files are provided open source on the Prusa Research website, and Prusa reports that at the time of writing, over 200,000 devices have been provided to healthcare workers in the Czech Republic. The Prusa PRO version has reportedly achieved compliance with European PPE regulations. The

University of Nebraska Medical Centre reports that these shields are easy to decontaminate after inoculation with bacterial colonies by using dilute bleach (Armijo et al. 2020). The Prusa devices have been printed for use in Brazil (Armijo et al. 2020), and over 10,000 units were printed within 5 weeks of commencing production in Paris (Lemarteleur et al. 2020). Indeed, even mainstream companies such as Ford Motors have commenced production of similar devices, and have reported production of around 1 million face shields per week via a combination of injection moulding and 3D printing from early in the pandemic (Ford Media Centre 2020). A group based in Frankfurt, Germany, has developed a skeletonised version of the shield which is faster to print, and uses less plastic than the Prusa version (Neijhoft et al. 2020). The authors note that when printed with PLA, warm water can be poured on the device to allow further manipulation of its shape to better fit individual wearers.

A huge array of modifications to the original designs have been produced specifically for dozens of different applications (Amin et al. 2020). One group in Mumbai, India, has added a binder clip to their device, intended for use during surgery, to secure the transparent PVC sheet in place in order to avoid the time-consuming process of punching holes in the plastic (Chaturvedi et al. 2020). However, they do not make an attempt to assess whether this increases the rate of accidental detachment of the sheet, which could be a serious issue in terms of loss of user protection and contamination of the sterile field during surgery. A further device combines a headlight holder with an attachment for a transparent protective face shield to protect otolaryngologists during patient examination (Viera-Artiles and Valdiande 2020). One interesting, but experimental method uses stereolithography 3D printing to form a hollow tube to which a transparent shield is attached. Compressed gas is then passed through the tube to flow from over the wearer's face, thus theoretically reducing air entrainment from the environment, and thus reducing infection risk (Maracaja et al. 2020).

3.2 3D Printed Masks

Surgical face masks primarily prevent droplet emissions by the wearer, although they also provide some protection against direct contact of droplets with the mucus membranes, as well as reducing the tendency of the user to touch their face. These masks are recommended for general public use, while the Centres for Disease Control and Prevention recommends the use of N95 masks when working with patients with COVID-19 (Centres for Disease Control and Prevention 2020). N95 masks are composed of a filter of much higher efficiency than surgical masks (designed to filter 95% of particles of 0.3 μm and greater in size), and must be fit tested to ensure that no particles are inhaled around the mask. Indeed, proper fit testing of the mask contributes much more to the efficiency of the mask than filter resolution (Grinshpun et al. 2009). A pre-pandemic study reported improved pressure on the wearer's face following an ABS printed attachment to the mask designed individually following LASER scanning of wearers' faces, although no formal fit testing was reported (Cai et al. 2018). Facial scanning has been used to produce a whole proof-of-concept face mask by selective laser sintering (Swennen et al. 2020). Printing of the device took around 12 hours, and allows for insertion of disposable filters and headbands. No clinical testing of the device was reported, with formal fit testing not undertaken (Scott et al. 2020). A similar device optimised for fused deposition modelling 3D printing, known as the NanoHack mask, is available from Copper3D (#HackThePandemic 2020).

Instead of dealing with the regulatory issues surrounding 3D printing of masks themselves, some makers have sought to increase the effectiveness of existing masks with 3D printed adjuncts. One group has designed a frame known as the Pan Fab to surround an N95 mask in order to replace the elastic bands (McAvoy et al. 2020). The pre-print study proposes that their device is likely to increase the durability of the masks while also improving the fit. The device's bendable metal wires between two 3D printed parts allow adjustment to fit multiple types of mask. Impressively, the group reports an increased fit test pass rate with the frame, although attempt no statistical analysis of their results. This device would indeed represent a helpful development in the field, improving the spectrum of masks available to healthcare workers who had previously failed a fit test.

Due to the ongoing problems of fit to wearers' faces, disinfection, production of filtration materials, and others, many researchers are investigating more sophisticated 3D printing techniques to manufacture PPE. The Stopgap Surgical Face Mask (SFM) Revision B is a 3D printable device allowing for custom printing of masks fitted to a user's face, and insertion of custom filters to suit different environments, although it does require specialist 3D printing techniques not available to many hobbyists (VHA Innovation 2020). Some have even investigated direct 3D printing of N95 filter polymers themselves (Ishack and Lipner 2020), although this is fraught with difficulty; even small defects in the material will render the mask useless, and strict regulation will no doubt be required for these novel materials. Another innovative design conceived by doctors working in Stanford, California, USA, quadruples the existing supply of N95 masks by cutting them into quarters, and using these quarters as filters in a bespoke 3D printed frame with a soft silicone printed base (Imbrie-Moore et al. 2020). Cut pieces of mask can be removed and replaced as necessary. While this solution is certainly inventive, sacrificing a tested and approved device to make four possibly inferior and untested devices is controversial. However, fit testing using a PortaCount device proved promising.

At least one hospital has published data regarding their experience with the use of 3D printed masks during the pandemic. Emergency PPE was provided to the orthopaedic team at a USA-based trauma centre due to shortages of conventional masks (Thomas et al. 2020). They used the Wiles COVID pandemic mask (ChrisWilesDO 2020) printed in PLA, with proprietary filter material (NanoNet and NanoForce Media by Cummins) found to be as effective as N95 filters. This material was chosen in part to avoid bottlenecks in the supply chain caused by the COVID-19 pandemic. Masks were customised with individualised names to boost morale and encourage reuse. However, production of the masks did involve a moderate amount of post-processing to finish edges in contact with the skin, and to manually cut and insert filters into the masks. A total of 50 units were produced, with a total production time estimated at approximately 350 hours. Printing in PLA allowed for heating of the mask edges and press-fitting to the face. While the design process is well described, the authors made no attempt to validate the effectiveness of their devices, or to quantify their actual usage.

3.3 3D Printed Powered Air-Purifying Respirators

PAPRs involve encasing the user's head within a hood, and use powered fans to entrain ambient air through a filter, thus providing filtered air to the wearer. They have a higher protection factor and lower breathing resistance than passive masks such as the N95, and often incorporate protection for the wearer's face and eyes. Erickson et al. have used a 3D printed manifold fastened over the usually open air intake fan to retrofit N95 filters to arthroplasty helmets left over following cancellation of elective surgeries in order to form a PAPR for use against COVID-19 (Erickson et al. 2020). They designed the part iteratively with feedback from the clinical team, thus highlighting one of the key strengths of 3D printing for this application. No significant CO_2 accumulation inside the hood was noted, and particle flow testing indicated protection equal to or better than that of a typical N95 mask. Hubbard et al. have created a similar device to repurpose USA-based firefighter respirators to work as PAPRs using 3D printed components (Hubbard and Pearce 2020). Their paper contains a detailed build log regarding the 3D printed mask adapter and powered filter backpack. The device performed well in the group's stringent tests to the National Institute for Occupational Safety and Health (NIOSH) standards, although it has not been officially certified. Some parts for existing PAPRs have even been printed using a novel technique for 3D scanning involving a medical CT scanner, resulting in impressive accuracy of product dimensions, although the devices remain untested at the time of writing (Coté et al. 2020). Non-powered versions of the devices have also caught the attention of makers, with several noticing the possibility of existing snorkel masks as a full-face PPE mask. These masks are then retrofitted with filters using a 3D printed adapter in order to provide filtration for the air intake (Thierry et al. 2020; Formlabs 2020). Similar devices have been tested via a standard fit test, as well as via end-tidal CO_2 monitoring, reportedly performing well (Germonpre et al. 2020).

3.4 Decontamination and Sterilisation of 3D Printed Materials

Decontamination and sterilisation of 3D printed materials has been shown to be effective. Plastics such as PA12, ABS, PLA, and SLA photopolymers may be decontaminated using 10% bleach, quaternary ammonium sanitiser, 3% hydrogen peroxide, and exposure to 70 °C heat for 30 minutes. All of these methods provided complete inactivation of SARS-CoV2 and a multitude of other pathogenic viruses under experimental conditions (Welch et al. 2020). Isopropyl alcohol 70% reportedly performed less well. Decontamination did not affect material integrity, although the group reported that PLA was permeable to viral particles, thus making it more difficult to decontaminate and less useful in the manufacture of medical devices. Post-processing measures may reduce the risk of bacterial contamination of 3D printed parts, with one group finding that coating PLA with acrylic acid reduced biofilm formation under experimental conditions (Muro-Fraguas et al. 2020). Sterilisation of 3D printed parts remains a problem due to their thermo-stable nature. One group has redesigned a RepRap-class 3D printer to facilitate enhanced extruder and bed temperatures (500 °C and 200 °C, respectively) to allow for printing of materials such as polyetherketoneketone and polyetherimide, a concept which was tested by the printing of the open-source Maker Mask (Maker Mask and Respitator 2020; Skrzypczak et al. 2020). Some authors have also suggested the integration of copper into 3D printed polymers in order to provide some antibacterial properties to the material (Zuniga and Cortes 2020). Despite these ongoing issues, the USA Food and Drug Administration in 2016 published their perspective on medical devices built using additive manufacturing (Di Prima et al. 2016). Interestingly, they state that the devices are classified by function rather than by method of manufacturing, unless there is a specific question of safety or effectiveness in relation to the method of manufacture. However, given the evidence above, care should be taken with porous 3D printing materials, especially when using PLA for production-grade devices.

4 Conclusion

The open-source community has made an inspiring contribution to the COVID-19 pandemic through the distributed manufacture of many tens of thousands of face shields to front-line workers across the globe. Masks remain a more difficult challenge, and formal fit testing is essential if the devices described in this chapter are to be used successfully in healthcare settings. One possible solution to the lack of masks is printing of filter cartridges to replace those attached to reusable masks, or to retrofit existing devices such as snorkel masks. This approach would eliminate the need for formal fit testing of every 3D printed device, and would bypass many of the regulatory issues; the printed part would not be directly in contact with the user's skin, but would instead simply hold the filter material in a cartridge attached to the mask. Despite the drawbacks mentioned above, the open-source community and makers throughout the globe, as well as the companies supporting them, have undoubtedly contributed greatly to the safety and morale of local front-line healthcare workers through their tireless contributions to the design, manufacture, and distribution of free and open-source personal protective equipment.

References

Amin D, Nguyen N, Roser SM, Abramowicz S (2020) 3D printing of face shields during COVID-19 pandemic: a technical note. J Oral Maxillofac Surg 78(8):1275–1278

Armijo PR, Markin NW, Nguyen S, Ho DH, Horseman TS, Lisco SJ, et al (2020) 3D printing of face shields to meet the immediate need for PPE in an anesthesiology department during the COVID-19 pandemic. Am J Infect Control

Burki T (2020) Global shortage of personal protective equipment. Lancet Infect Dis 20(7):785–786

Cai M, Li H, Shen S, Wang Y, Yang Q (2018) Customized design and 3D printing of face seal for an N95 filtering facepiece respirator. J Occup Environ Hyg 15(3):226–234

Centres for Disease Control and Prevention (2019) Interim domestic guidance on the use of respirators to prevent transmission of SARS. Centres for Disease Control and Prevention, Atlanta, Georgia, USA. https://www.cdc.gov/sars/clinical/respirators.html

Centres for Disease Control and Prevention (2020) Operational Considerations for Personal Protective Equipment in the Context of Global Supply Shortages for Coronavirus Disease 2019 (COVID-19) Pandemic: non-US Healthcare Settings. Centres for Disease Control and Prevention, Atlanta, Georgia, USA. https://www.cdc.gov/coronavirus/2019-ncov/hcp/non-us-settings/emergency-considerations-ppe.html

Chan JF-W, Yuan S, Kok K-H, To KK-W, Chu H, Yang J et al (2020) A familial cluster of pneumonia associated with the 2019 novel coronavirus indicating person-to-person transmission: a study of a family cluster. Lancet 395(10223):514–523

Chaturvedi S, Gupta A, Krishnan SV, Bhat AK (2020) Design, usage and review of a cost effective and innovative face shield in a tertiary care teaching hospital during COVID-19 pandemic. J Orthop 21:331–336

Cheung JC-H, Ho LT, Cheng JV, Cham EYK, Lam KN (2020) Staff safety during emergency airway management for COVID-19 in Hong Kong. Lancet Respir Med 8(4):e19

ChrisWilesDO (2020) Wiles covid pandemic mask. NIH 3D print exchange. https://3dprint.nih.gov/discover/3dpx-013805

Chu DK, Akl EA, Duda S, Solo K, Yaacoub S, Schünemann HJ et al (2020) Physical distancing, face masks, and eye protection to prevent person-to-person transmission of SARS-CoV-2 and COVID-19: a systematic review and meta-analysis. Lancet 395(10242):1973–1987

Cook TM (2020) Personal protective equipment during the coronavirus disease (COVID) 2019 pandemic—a narrative review. Anaesthesia 75(7):920–927

Coté JJ, Haggstrom J, Vivekanandan R, Coté KA, Real DL, Weber DP, et al (2020) COVID-19 and a novel initiative to improve safety by 3D printing personal protective equipment parts from computed tomography. 3D Print Med 6(1):20

ctwiles (2020) Mask Frame CEG Extreme and small n95 fitter (High Filtration with Halyard H600). Thingiverse database, MakerBot Industries, Brooklyn, NY, USA. Published: 05 Apr 2020. https://www.thingiverse.com/thing:4262131

D4Care (2020) 3D4Care, Paris, France

de Araujo Gomes B, Queiroz FLC, Pereira PL de O, Barbosa TV, Tramontana MB, Afonso FAC, et al (2020) In-house three-dimensional printing workflow for face shield during COVID-19 pandemic. J Craniofac Surg 31(6):e652–e653

Di Prima M, Coburn J, Hwang D, Kelly J, Khairuzzaman A, Ricles L (2016) Additively manufactured medical products—the FDA perspective. 3D Print Med 2

Erickson MM, Richardson ES, Hernandez NM, Bobbert DW, Gall K, Fearis P (2020) Helmet modification to PPE with 3D printing during the COVID-19 pandemic at Duke University medical center: a novel technique. J Arthroplasty 35(7):S23–S27

Feinmann J (2020) PPE: what now for the global supply chain? BMJ 15:m1910

Flanagan ST, Ballard DH (2020) 3D Printed face shields: a community response to the COVID-19 global Pandemic. Acad Radiol 27(6):905–906

Ford Media Centre (2020). Ford is making face masks and face shields to enable employees across Europe to return safely to work. Ford Media Centre, Ford Motor Company, Dearborn, Michigan,

United States of America. https://media.ford.com/content/fordmedia/feu/en/news/2020/04/30/ford-is-making-face-masks-and-face-shields-to-enable-employees-a.html

Formlabs (2020) 3D printed adapters for Decathlon snorkel mask conversion to emergency PPE. Formlabs, Somerville, Massachusetts, United States of America. https://formlabs.com/uk/covid-19-response/snorkel-mask-emergency-ppe-adapters/

François P-M, Bonnet X, Kosior J, Adam J, Khonsari RH (2020) 3D-printed contact-free devices designed and dispatched against the COVID-19 pandemic: the 3D COVID initiative. J Stomatol Oral Maxillofac Surg

Frazer JS, Tyrynis Everden AJ (2020) Emerging patterns of hypercoagulability associated with critical COVID-19: a review. Trends Anaesth Crit Care S2210844020300903

Frazer JS, Shard A, Herdman J (2020) Involvement of the open-source community in combating the worldwide COVID-19 pandemic: a review. J Med Eng Technol 44(4):169–176

Germonpre P, Van Rompaey D, Balestra C (2020) Evaluation of protection level, respiratory safety, and practical aspects of commercially available snorkel masks as personal protection devices against aerosolized contaminants and SARS-CoV2. Int J Environ Res Public Health 17(12)

Global Construction Review (2020) Instant isolation rooms: Chinese firm prints quarantine pods for hospitals struggling with virus. Global construction review, Chartered Institute of Building, Atom, London, United Kingdom. Published: 13 Mar 2020. https://www.globalconstructionreview.com/news/instant-isolation-rooms-chinese-firm-prints-quaran/

Griffiths L, O'Connor D, Davies S (2020) The latest 3D printing efforts against Covid-19. TCT Magazine, Rapid News Publications Ltd. https://www.tctmagazine.com/additive-manufacturing-3d-printing-news/live-blog-how-the-3d-printing-industry-fighting-covid-19/

Grinshpun SA, Haruta H, Eninger RM, Reponen T, McKay RT, Lee S-A (2009) Performance of an N95 filtering facepiece particulate respirator and a surgical mask during human breathing: two pathways for particle penetration. J Occup Environ Hyg 6(10):593–603

#HackThePandemic (2020) Copper 3D. https://copper3d.com/hackthepandemic/

Hamner L, Dubbel P, Capron I, Ross A, Jordan A, Lee J et al (2020) High SARS-CoV-2 attack rate following exposure at a choir practice—Skagit county, Washington. MMWR Morb Mortal Wkly Rep 69(19):606–610

Hubbard BR, Pearce JM (2020) Conversion of self-contained breathing apparatus mask to open source powered air-purifying particulate respirator for fire fighter COVID-19 response. HardwareX 8:e00129

Interesting Engineering (2020) Boy Scout 3D prints numerous "Ear Guards" to help medical staff suffering from face masks. Interesting Engineering, Inc. Published: 07 Apr 2020. https://interestingengineering.com/boy-scout-3d-prints-numerous-ear-guards-to-help-medical-staff-suffering-from-face-masks

Imbrie-Moore AM, Park MH, Zhu Y, Paulsen MJ, Wang H, Woo YJ (2020) Quadrupling the N95 supply during the COVID-19 crisis with an innovative 3D-printed mask adaptor. Healthcare 8(3):225

Ishack S, Lipner SR (2020) Applications of 3D printing technology to address COVID-19-related supply shortages. Am J Med 133(7):771–773

Jayaweera M, Perera H, Gunawardana B, Manatunge J (2020) Transmission of COVID-19 virus by droplets and aerosols: a critical review on the unresolved dichotomy. Environ Res 188:109819

Johns Hopkins University of Medicine Coronavirus Resource Centre (2020) Johns Hopkins University of Medicine, Baltimore, MD, USA. https://coronavirus.jhu.edu/map.html

Lemarteleur V, Fouquet V, Le Goff S, Tapie L, Morenton P, Benoit A, et al (2020) 3D-printed protected face shields for health care workers in Covid-19 pandemic. Am J Infect Control

Lindsley WG, Noti JD, Blachere FM, Szalajda JV, Beezhold DH (2014) Efficacy of face shields against cough aerosol droplets from a cough simulator. J Occup Environ Hyg 11(8):509–518

Liu Y, Ning Z, Chen Y, Guo M, Liu Y, Gali NK et al (2020) Aerodynamic analysis of SARS-CoV-2 in two Wuhan hospitals. Nature 582(7813):557–560

Lu C-W, Liu X-F, Jia Z-F (2020) 2019-nCoV transmission through the ocular surface must not be ignored. Lancet 395(10224):e39

Ma J, Qi X, Chen H, Li X, Zhan Z, Wang H, et al (2020) Exhaled breath is a significant source of SARS-CoV-2 emission. Public and Global Health. http://medrxiv.org/lookup/doi/10.1101/2020.05.31.20115154
Maker Mask, 3D Printable Respitator (2020) Maker Mask, Seattle, Washington, United States of America. https://www.makermask.com/
Maracaja L, Blitz D, Maracaja DLV, Walker CA (2020) How 3D printing can prevent spread of COVID-19 among healthcare professionals during times of critical shortage of protective personal equipment. J Cardiothorac Vasc Anesth 34(10):2847–2849
McAvoy M, Bui A-TN, Hansen C, Plana D, Said JT, Yu Z, et al (2020) 3D printed frames to enable reuse and improve the fit of N95 and KN95 respirators. medRxiv: the preprint server for health sciences
Morawska L, Cao J (2020) Airborne transmission of SARS-CoV-2: the world should face the reality. Environ Int 139:105730
Morawska L, Milton DK (2020) It is time to address airborne transmission of coronavirus disease 2019 (COVID-19). Clin Infect Dis 6:ciaa939
Morawska L, Tang JW, Bahnfleth W, Bluyssen PM, Boerstra A, Buonanno G et al (2020) How can airborne transmission of COVID-19 indoors be minimised? Environ Int 142:105832
Muro-Fraguas I, Sainz-García A, López M, Rojo-Bezares B, Múgica-Vidal R, Sainz-García E et al (2020) Antibiofilm coatings through atmospheric pressure plasma for 3D printed surgical instruments. Surf Coat Technol 15(399):126163
Neijhoft J, Viertmann T, Meier S, Söhling N, Wicker S, Henrich D et al (2020) Manufacturing and supply of face shields in hospital operation in case of unclear and confirmed COVID-19 infection status of patients. Eur J Trauma Emerg Surg 46(4):743–745
Nicas M, Nazaroff WW, Hubbard A (2005) Toward understanding the risk of secondary airborne infection: emission of respirable pathogens. J Occup Environ Hyg 2(3):143–154
Ofner-Agostini M, Gravel D, McDonald LC, Lem M, Sarwal S, McGeer A et al (2006) Cluster of cases of severe acute respiratory syndrome among Toronto healthcare workers after implementation of infection control precautions: a case series. Infect Control Hosp Epidemiol 27(5):473–478
Peiris JSM, Yuen KY, Osterhaus ADME, Stöhr K (2003) The severe acute respiratory syndrome. N Engl J Med 349(25):2431–2441
Prusa Research (2020) 3D printed face shields for medics and professionals. Prusa Research, Prague, Czech Republic. https://www.prusa3d.com/covid19/
RCLifeOn (2017) Webbed glove. Thingiverse database, MakerBot Industries, Brooklyn, NY, USA. Published: 06 July 2017. https://www.thingiverse.com/thing:4262131
Roberge RJ (2016) Face shields for infection control: a review. J Occup Environ Hyg 13(4):235–242
Ronco C, Reis T, Husain-Syed F (2020) Management of acute kidney injury in patients with COVID-19. Lancet Respir Med 8(7):738–742
Sapoval M, Gaultier AL, Del Giudice C, Pellerin O, Kassis-Chikhani N, Lemarteleur V et al (2020) 3D-printed face protective shield in interventional radiology: evaluation of an immediate solution in the era of COVID-19 pandemic. Diagn Interv Imaging 101(6):413–415
Scott AR, Hu J, Gan C, Morris JA, Meacham KW, Ballard DH (2020) Safety concerns for facial topography customized 3D-printed N95 filtering face-piece respirator produced for the COVID-19 pandemic: initial step is respiratory fit testing. Int J Oral Maxillofac Surg S0901502720303337
Seah I, Agrawal R (2020) Can the coronavirus disease 2019 (COVID-19) affect the eyes? a review of coronaviruses and ocular implications in humans and animals. Ocul Immunol Inflamm 28(3):391–395
Seah IYJ, Anderson DE, Kang AEZ, Wang L, Rao P, Young BE et al (2020) Assessing viral shedding and infectivity of tears in coronavirus disease 2019 (COVID-19) patients. Ophthalmology 127(7):977–979
Siegel JD, Rhinehart E, Jackson M, Chiarello L (2007) Health care infection control practices advisory committee. 2007 guideline for isolation precautions: preventing transmission of infectious agents in health care settings. Am J Infect Control 35(10 Suppl 2):S65–S164

Siripanthong B, Nazarian S, Muser D, Deo R, Santangeli P, Khanji MY et al (2020) Recognizing COVID-19-related myocarditis: the possible pathophysiology and proposed guideline for diagnosis and management. Heart Rhythm 17(9):1463–1471

Skrzypczak NG, Tanikella NG, Pearce JM (2020) Open source high-temperature RepRap for 3-D printing heat-sterilizable PPE and other applications. HardwareX 8:e00130

Stadnytskyi V, Bax CE, Bax A, Anfinrud P (2020) The airborne lifetime of small speech droplets and their potential importance in SARS-CoV-2 transmission. Proc Natl Acad Sci USA 117(22):11875–11877

Swennen GRJ, Pottel L, Haers PE (2020) Custom-made 3D-printed face masks in case of pandemic crisis situations with a lack of commercially available FFP2/3 masks. Int J Oral Maxillofac Surg 49(5):673–677

Tarfaoui M, Nachtane M, Goda I, Qureshi Y, Benyahia H (2020) Additive manufacturing in fighting against novel coronavirus COVID-19. Int J Adv Manuf Technol 17:1–15

Thierry B, Célérier C, Simon F, Lacroix C, Khonsari R-H (2020) How and why use the EasyBreath® surface snorkeling mask as a personal protective equipment during the COVID-19 pandemic? Eur Ann Otorhinolaryngol Head Neck Dis 137(4):329–331

Thomas RJ (2013) Particle size and pathogenicity in the respiratory tract. Virulence 4(8):847–858

Thomas CN, Schroder LK, Cole PA (2020) Ten days to implementation of 3D-printed masks for a level-i orthopaedic trauma practice during the COVID-19 pandemic. J Bone Joint Surg Am 102(16):e95

Tran K, Cimon K, Severn M, Pessoa-Silva CL, Conly J (2012) Aerosol generating procedures and risk of transmission of acute respiratory infections to healthcare workers: a systematic review. PLoS ONE 7(4):e35797

van Doremalen N, Bushmaker T, Morris DH, Holbrook MG, Gamble A, Williamson BN, et al (2020) Aerosol and surface stability of SARS-CoV-2 as compared with SARS-CoV-1. N Engl J Med 382(16):1564–1567

VHA Innovation (2020) Stopgap surgical face mask (SFM) Revision B. NIH 3D print exchange. https://3dprint.nih.gov/discover/3dpx-014168

Viera-Artiles J, Valdiande JJ (2020) 3D-printable headlight face shield adapter. personal protective equipment in the COVID-19 era. Am J Otolaryngol 41(5):102576

Wang D, Hu B, Hu C (2020) Clinical characteristics of 138 hospitalised patients with 2019 novel coronavirus-infected pneumonia in Wuhan. China JAMA 323:1061–1069

Welch JL, Xiang J, Mackin SR, Perlman S, Thorne P, O'Shaughnessy P et al (2020) Inactivation of severe acute respiratory coronavirus virus 2 (SARS-CoV-2) and diverse RNA and DNA viruses on three-dimensionally printed surgical mask materials. Infect Control Hosp Epidemiol 12:1–8

Wesemann C, Pieralli S, Fretwurst T, Nold J, Nelson K, Schmelzeisen R, et al (2020) 3-D printed protective equipment during COVID-19 Pandemic. Materials (Basel, Switzerland) 13(8)

World Health Organisation (2020a) Consensus document on the epidemiology of severe acure respiratory syndrome (SARS). Department of Communicable Disease Surveillance and Response, World Health Organization, Geneva, Switzerland. Published: 2003. https://apps.who.int/iris/bitstream/handle/10665/70863/WHO_CDS_CSR_GAR_2003.11_eng.pdf

World Health Organisation (2020b) Shortage of personal protective equipment endangering health workers worldwide. World Health Organization, Geneva, Switzerland. Published: 03 Mar 2020. https://www.who.int/news/item/03-03-2020-shortage-of-personal-protective-equipment-endangering-health-workers-worldwide

World Health Organization (2020c). Summary of probable SARS cases with onset of illness from 1 Nov 2002 to 31 July 2003. World Health Organization, Geneva, Switzerland. Published 2004. http://www.who.int/csr/sars/country/table2004_04_21/en/

Wu P, Duan F, Luo C, Liu Q, Qu X, Liang L et al (2020) Characteristics of ocular findings of patients with coronavirus disease 2019 (COVID-19) in Hubei province. China JAMA Ophthalmol 138(5):575–578

Yu ITS, Li Y, Wong TW, Tam W, Chan AT, Lee JHW et al (2004) Evidence of airborne transmission of the severe acute respiratory syndrome virus. N Engl J Med 350(17):1731–1739

Zeng W, Wang X, Li J, Yang Y, Qiu X, Song P, et al (2020) Association of daily wear of eyeglasses with susceptibility to coronavirus disease 2019 infection. JAMA Ophthalmol
Zuniga JM, Cortes A (2020) The role of additive manufacturing and antimicrobial polymers in the COVID-19 pandemic. Expert Rev Med Devices 17(6):477–481

Effectiveness of 3D Printing and Open-Source Technologies for Development of Ventilators, and Other Critical Care Technology in the Context of the COVID-19 Pandemic

John Scott Frazer

Abstract The unprecedented global COVID-19 pandemic has fuelled an explosion of attempts to manufacture open-source ventilators, diagnostic equipment, and personal protective equipment by both individuals and companies. The initial and well-publicised success of Insinnova's 3D printed venturi ventilator valve to boost supply, as well as global panic regarding the availability of intensive care unit ventilators, has led the open-source community to focus on this area when considering projects. However, patients with acute respiratory distress syndrome such as that caused by COVID-19 are notoriously difficult to ventilate, with clinical experience suggesting that even slight adjustments in ventilator pressures or modes can lead to a dramatic deterioration in already profoundly unstable patients. In the following chapter, we review strategies for the production of open-source and 3D printed solutions to critical care technology in response to the pandemic. We focus on open-source ventilators in this chapter, and discuss the positive and negative implications of open-source ventilator designs on clinical management of patients, as well as the highly discouraging prospect of ventilator splitting for management of multiple patients with a single ventilator. We discuss possible further opportunities in critical care technology for makers to explore, including haemofiltration devices, and infusion pumps.

Keywords Ventilators · Human care · COVID-19

Abbreviations

FDM	Fused Deposition Modelling
SLA	Stereolithography
ABS	Acrylonitrile Butadiene Styrene

J. S. Frazer (✉)
Somerville College, University of Oxford, Oxford, UK
e-mail: drjsfrazer@gmail.com

K. Sandhu et al. (eds.), *Emerging Applications of 3D Printing During CoVID 19 Pandemic*, Lecture Notes in Bioengineering, https://doi.org/10.1007/978-981-33-6703-6_3

PLA Polylactic Acid
BVM Bag Valve Mask
PEEP Positive End-Expiratory Pressure

1 Introduction

While COVID-19 has far-reaching consequences throughout the entire body, its effects on the lungs are particularly marked. Acute respiratory distress syndrome (ARDS) is the most feared complication, with a 28-day mortality of up to 44% (Sinha et al. 2020). Pandemic modelling has resulted in predictions of rapid overwhelming of critical care infrastructure, with a global shortfall of 880,000 ventilators anticipated in an influenza pandemic by some sources (GlobalData 2020). Many have warned of difficult triage processes due to lack of intensive care unit (ICU) capacity (Truog et al. 2020). As a result, there has been incredible support from the open-source community in the production of materials and design of devices to fight the pandemic and support local healthcare systems.

Early in a pandemic surge situation, ICUs do have some capacity to expand; unlike many other departments, most anaesthetists have had some experience in intensive care medicine. When elective theatre lists are cancelled, anaesthetists are able to take on some roles in ICU such as daily clinical review of patients, adjustment of ventilator parameters, insertion of lines, and endotracheal intubation. Theatre recovery and other clinical areas may also be used as a temporary ICU; during the COVID-19 pandemic, some UK ICUs were forced to increase capacity by over 200% (Chong 2020). However, as the pandemic intensified, and as models predicted massive overwhelming of even this increased capacity, anticipated ventilator shortages were widely publicised. Indeed, shortages of all critical care equipment, such as syringe pumps, haemofiltration circuits, and even medications commonly utilised in critical care, as well as expert nursing staff, was widespread. Lower national income exacerbates this trend; during the 2009 influenza pandemic, high-income countries had a lower ICU death toll than high-middle income and low-middle income countries (Duggal et al. 2016).

A number of early successes in critical care and ventilator technology under other circumstances have inspired work in this area. The intensive care unit was borne out of the 1952 epidemic of polio, a viral infection that invades nerves and results in paralysis (Lassen 1953). Paralysis of respiratory muscles due to polio or other causes results in a condition known as type 2 respiratory failure, which causes a decrease in the arterial concentration of oxygen and an increase in carbon dioxide. Hundreds of patients throughout Europe were afflicted by respiratory failure caused by polio, and many owe their lives to the 'iron lungs' invented during this time to provide ventilation. These devices were comparatively simple machines by modern standards, and were designed to replace the function of the denervated respiratory muscles using negative pressure to force gas into and out of the lungs of a patient encased in the machine. Modern and emotive stories exist of patients with Guillain–Barre syndrome, a reversible condition resulting in temporary paralysis caused by a viral infection, ventilated manually by relatives for hours or days (deBoisblanc 2005; Chandler 2020). During the current COVID-19 pandemic, the early success of the Italian company Isinnova, who reverse engineered and 3D printed a ventilator venturi mixing valve for Italian hospitals, captivated the imagination of makers throughout the world (Kleinman 2020). Attention has now turned to the vastly more complex issue of designing mechanical ventilators for critically unwell patients.

This chapter commences with an overview of the pathophysiology of COVID-19 ARDS, and the clinical and logistical challenges these patients present to critical care facilities. We provide a description of the key operational characteristics of a critical care ventilator, followed by an investigation of the response of the open-source community and industry to the proposed shortage of ventilators, highlighting points at which 3D printing and other rapid prototyping methods have been utilised to greatest effect. In this chapter, we catalogue the multitude of designs made available in response to the pandemic, comparing them to the ideal specifications of a critical care ventilator and exploring their utility in managing patients with COVID-19. Further to this, we propose options for future design and research of critical care technology to combat COVID-19.

2 COVID-19 Pathophysiology

The clinical syndrome termed COVID-19 is caused by the SARS-CoV2 virus, related to the pathogen associated with severe acute respiratory syndrome (SARS) and Middle East respiratory syndrome (MERS). Many cases of COVID-19 are asymptomatic, although a significant although as yet undetermined proportion go on to develop a dangerous form of lung injury known as ARDS, as well as systemic effects such as myocarditis (Siripanthong et al. 2020), acute kidney injury (Ronco et al. 2020), and hypercoagulability (Frazer and Everden 2020), thus leading to multi-organ failure. The 2012 Berlin Criteria define ARDS as a clinical condition affecting the lungs causing severe hypoxaemia (Force et al. 2012). The condition is likely to also include bilateral opacities on chest radiographs, reduced lung compliance, and a requirement for high-positive end-expiratory pressure (PEEP) and minute volume. In contrast to type 2 respiratory failure, which occurs as a result of the inability to move gas into and out of the lungs, ARDS results from direct damage to the gas exchange mechanism of the lungs and causes type 1 respiratory failure, resulting in a reduction of arterial oxygen concentration without affecting the carbon dioxide concentration in the early stages. ARDS has many causes, classified as those having a direct effect on the lungs (mechanical ventilation, inhalational lung injury, traumatic lung contusions, drowning, aspiration, pneumonia, and others), and those whose effects are indirect (shock, sepsis, embolism, burns, raised intracranial pressure, transfusion-associated lung injury, and others). The direct causes are more likely to be associated with epithelial damage (i.e. the alveolar surface in contact with the air), whereas indirect causes are associated with endothelial damage (i.e. the lining of the capillaries in contact with the blood) (Shaver and Bastarache 2014). Patients with COVID-19 are increasingly thought to have an atypical form of ARDS, with some evidence that this is due to a form of hypercoagulability confined specifically to the pulmonary vasculature (Frazer and Everden 2020), although this remains an area of intense debate. Whatever the cause, patients with ARDS are notoriously unstable, and a number of ventilatory strategies have been trialled in order to attempt to optimise outcomes in this high-mortality group.

When considering the design of open-source ventilators to combat COVID-19, intimate knowledge of the respiratory mechanics of ARDS is required. The refractory hypoxaemia experienced by this group of patients compounds problems with every body system and impairs healing. Progression to type 2 respiratory failure is a concerning and late finding. The high tidal volumes and pressures used in the early days of ARDS management resulted in normalisation of arterial carbon dioxide concentration, but also caused barotrauma. Later, a strategy of tolerating hypercapnia in exchange for reduced pressures and tidal volumes was adopted and appears to have reduced this complication (Matthay et al. 2019). Thus, precise control of respiratory function is crucial to the stability and improvement of patients with ARDS, with urgent and drastic alterations to ventilatory mechanics often sparing patients from death. Several key facets in the management of ARDS in critical care have been identified: lung-protective ventilation (Brower et al. 2000); prone positioning; negative fluid balance; aggressive infection control; adequate sedation and neuromuscular blockade. Indeed, ventilatory strategies for the management of patients with COVID-19 are still under intense debate. These ongoing discussions highlight the need for sophisticated ventilators with advanced measurement of respiratory parameters, not only in order to treat patients suffering from COVID-19 associated ARDS, but to gather data to further elucidate the respiratory mechanics of the illness. Indeed, many authors highlight the critical need for ventilators that sense tidal volume, inspiratory pressure, intrinsic PEEP, and other values, mandating that simple 'transport ventilator' style devices are inappropriate for the management of patients with COVID-19 (Gruber et al. 2006).

3 Operational Characteristics of Critical Care Ventilators

At the most basic level, a standard positive pressure mechanical ventilator incorporates a breathing circuit connected to the patient's lungs, and a device to deliver alternating high and low pressure to facilitate the movement of gas into and out of the circuit. This most basic ventilator concept will have a setting for the pressure maintained when the patient is in expiration (termed the positive end-expiratory pressure; PEEP), inspiration, and a setting for the timing of cycling between the two (the respiratory rate). Ideally, the machine should be capable of altering the time spent in inspiration and expiration, typically expressed as the inspiratory to expiratory

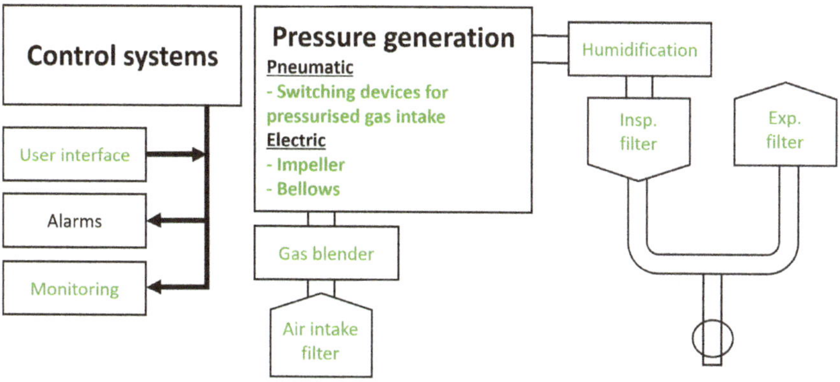

Fig. 1 Simple schematic diagram to illustrate the various components of a simple ventilator. Components partially or fully amenable to 3D printing are highlighted in green. Note that this is not an exhaustive or complete diagram

(I:E) ratio. Control of the fraction of inspired oxygen (FiO_2) is a key feature of any ventilator which is to be used in ARDS due to the ongoing and often severe hypoxaemia experienced by these patients. A schematic diagram of a simple critical care ventilator is shown in Fig. 1, with parts most amenable to 3D printing highlighted.

A multitude of sensors must also be incorporated into any ventilator in order to facilitate safe ventilation. Facilities to measure or calculate tidal volume, accurately sense pressures, and provide information about circuit disconnections are mandatory. When a patient is pharmacologically paralysed or deeply sedated, the ventilator should deliver mandatory breaths to the patient. When the patient is more awake and taking spontaneous breaths, the ventilator should sense this attempt to take a breath, and deliver pressure to support this, seamlessly transitioning between the two modes without losing PEEP at any stage. Simple ventilators may not deliver a breath when the patient attempts to breathe in, or may deliver a mandatory breath when the patient breathes out or coughs; both will cause barotrauma to the lungs and could prove fatal in a patient with already critically damaged lungs. Alarms are another key feature of any usable ventilator. Alarms for any major unintentional change in machine state are mandatory, alerting operators to issues such as supply failure, accidental ventilation pause, high inspiratory pressure, loss of PEEP or inspiratory pressure, and derangement of tidal volume. Continuous display of accurate and up-to-date ventilatory parameters is crucial.

While some design specifications have been developed in order to better circumscribe the problem of ventilator design during the COVID-19 pandemic, many devices are not designed by individuals with any clinical experience of ventilation, or indeed of the incredibly specialised scenario of ventilating a patient suffering from ARDS. The Massachusetts Institute of Technology (MIT), which has developed a number of open-source ventilator solutions, lay out a series of basic design criteria (MIT 2020). They advocate requirements for adjustment of tidal volume, respiratory rate, I:E ratio, trigger sensitivity for supportive modes, maximum plateau pressure, and others. Even this design specification has a key flaw, however, stating that "Ventilation on room air is better than no ventilation at all. Blending of oxygen and air-gas mixture to adjust FiO_2 is not important in an emergency scenario" (MIT 2020). This statement overlooks the fact that patients suffering from type 1 respiratory failure due to ARDS are intubated and ventilated primarily due to direct damage to the lungs and the resulting hypoxia, or severe respiratory tiring due to hypoxia, rather than a simple failure of the muscles of ventilation due to paralysis. Thus, accurate control of FiO_2, and maintenance of high FiO_2, is crucial in the management of patients with severe ARDS. In this case, ventilating the patient with room air is not acceptable, and will likely cause more damage than if the patient were left to self-ventilate on supplemental oxygen therapy, if it does not first lead to hypoxic cardiac arrest and death. Indeed, simply taking patients off oxygen in order to facilitate endotracheal intubation has resulted in almost immediate desaturation and in some cases hypotension and cardiac arrest (Yao et al. 2020). Thus, given the risks of hypoxia, any ventilator designed for a respiratory pandemic must non-negotiably have a robust and simple mechanism of adjusting FiO_2, and ideally an option to provide a FiO_2 of 100% for short periods to oxygenate the patient during acute desaturations.

The UK Government Medicines and Healthcare products Regulatory Agency (MHRA) rapidly manufactured ventilator system guidelines make for harrowing reading for the hobbyist ventilator designer (Medicines and Healthcare products Regulatory Agency 2020). The document is an essential primer for anyone interested in the field, laying out the vital components of a pandemic ventilator, including strict pressure tolerances, safety features, battery backup, and even absence of electrical interference which may play havoc on other critical care machinery (a feature commonly overlooked amongst the hobbyist designs). In addition to this, the guidelines sensibly suggest that a provision for maintenance of pressure during in-line suctioning for removal of secretions is incorporated into ventilator designs, and that spontaneous ventilation modes are highly desirable as patients continue to wean from ventilation. Certain alarms are mandatory, as is continuous breath-by-breath monitoring and display of parameters. They stipulate continuous operation (100% duty cycle) for 14 days.

4 Open-Source Ventilators

The current COVID-19 pandemic has pushed critical care to the forefront of the public mind. While previous pandemics have sparked similar fears of equipment and healthcare shortages, these pandemics did not have the benefit of a dedicated open-source community with sophisticated technology at its disposal. The earliest open-source ventilator came in the wake of the 2003 H5N1 pandemic, with fear of a resurgence in 2008 sparking the creation of the Pandemic Ventilator (Pandemic Ventilator Project 2020). This design relied upon a compressed gas source, with no internal method of gas compression described in the design schematics, and no way of mixing gases to alter the FiO_2. This early prototype is remarkable for the use of a programmable logic controller, which is more commonly seen in industry than in the maker sphere. The creation of the Arduino microcontroller in 2005 has since sparked widespread use of microcontrollers in open-source electronics projects, and bridged the gap between prototype and manufacture, with the ability to integrate similar microcontroller architectures directly onto printed circuit boards. Ease of use and rapid iterations of firmware also lends itself to effective design and testing of what is the most complex and critical part of ventilator design, the control system.

A list of open-source ventilators is currently maintained on GitHub (Covid19-vent-list 2020). An informal pandemic ventilator scoring system has been devised, with each design scored from 0 to 5 for openness, buildability, community support, functionality testing, reliability testing, COVID-19 suitability, and clinician friendliness (Read 2020). In the sections below, we further classify types of open-source ventilators by design, assessing the strengths and limitations of each in turn. Table 1 summarises the open-source ventilators most amenable to 3D printing.

4.1 Bag Valve Mask Designs

A bag valve mask (BVM) is a hand-held bag which is at one end connects to a mask or breathing circuit attached to the patient, and at the other end connects to an oxygen supply, with a one-way valve between. The bag is manually squeezed to deliver a breath to the patient with exhalation occurring to the air. Most commonly used emergency BVMs are self-inflating, allowing ventilation even without a pressurised gas supply. Anaesthetic BVMs rely on a pressurised gas supply to inflate, although they allow more precise control of the breaths delivered to the patient. These devices may be combined with a PEEP valve to provide expiratory resistance. BVMs are not designed for long-term ventilatory support (Petsiuk et al. 2020), and patients are typically transferred from the BVM to a mechanical ventilator as rapidly as possible if this is required. However, because of their widespread availability and comparatively simple design, these devices have been the focus of much of the open-source effort for use as a bellows. Most of the hobbyist ventilators rely on a BVM operated by an actuator (Pearce 2020), many of which are designed specifically to be 3D printable from freely available designs (Petsiuk et al. 2020; Carlota 2020).

Although a negative pressure design using a Waters circuit exists (Alsema 2020), most makers have chosen the more widely available option of the self-inflating BVM to simplify mechanics. Some early systems used a motorised belt system to compress the bag (Costa 2020; OpenVent 2020), although these were prone to wear, jams, and incomplete emptying of the bag (Al Husseini et al. 2010). The current most common design involves compression of the BVM between two paddles (Hewing et al. 2020; Monolithic Power Systems 2020), which can be actuated by a lead screw (OpenBuilds 2020; joelfrax 2020), bespoke linear actuators (ProgressTH 2020; Rice University 2020; University of Minnesota Medical School 2020), or gears (zzzrrrzz 2020; grumpytacos 2020), although various types of designs have been attempted (AmboVent 2020). Some makers have used car windscreen wiper motors to drive

Table 1 List of characteristics of 3D ventilators described in this chapter. It should be noted that this is by no means an exhaustive list

Class of ventilator	Device	Printing method	Material	Comments	References
BVM	Petsiuk design	FDM	PLA + Ninjaflex	3D printed housing, gear assembly, control panel, and some pneumatic *components*	Petsiuk et al. (2020)
	Leitat technology centre design	FDM	ABS/PLA	3D printed mechanism, now undergoing industrial production in Spain. Reportedly has been tested on animals for 6 hours, and patients with COVID-19 for 2 hours, following laboratory tests over 48 hours	Carlota (2020, Leitat Managing Technologies (2020a, b)
	Mechanic ventilator, by Costa_D	FDM	ABS/PLA	Uses belt to compress bag	Costa (2020)
	OpenVent	FDM	ABS/PLA	Belt-driven arm to compress bag	OpenVent (2020)
	MPS Open-Source Ventilator	FDM	ABS/PLA	Paddle compression design	Monolithic Power Systems (2020)
	V-slot ventilator	FDM	ABS/PLA	Paddle compression design actuated by lead screw	OpenBuilds (2020)
	Frax3D ventilator 2.0—Ambu	FDM	ABS/PLA	Paddle compression design actuated by lead screw	joelfrax (2020)
	Progress TH design	FDM	ABS/PLA	Fully 3D printed self-contained paddle and gear design	ProgressTH (2020)
	Rice University ApolloBVM design	FDM	ABS/PLA	Mostly LASER cut design, with 3D printed gear and paddle components	Rice University (2020)
	Emergency Ventilator (EV-02)	FDM	ABS/PLA	Geared mechanism for actuating paddles. Printed housing and control panel	zzzrrrzz (2020)
	Grumpytacos design	FDM	ABS/PLA	Geared mechanism for actuating paddles. Printed housing	grumpytacos (2020)
	AmboVent	FDM	ABS/PLA	Designed to have a 3D printable housing, although no attempts at 3D printing recording by developers	AmboVent (2020)
	ThumperX	FDM	ABS/PLA	Lever arm design with swappable drill battery	ThumperX (2020)
	Ad hoc ventilator	3D printer as ventilator		3D printer itself used as ventilator	Ad Hoc Ventilator (2020)

Table 1 (continued)

Class of ventilator	Device	Printing method	Material	Comments	References
	3DPaV (3D-printer-as-ventilator)	3D printer as ventilator		3D printer itself used as ventilator	keeevvviiinnn (2020)
	Dhanani design	FDM	ABS/PLA with gears eventually replaced with nylon	Used CNC cut plywood plus ABS and PLA gears. Of note, these gears softened during testing on porcine models with high PEEP and were eventually replaced with nylon. Manufactured in just two hours	Dhanani et al. (2020)
Compressor/ impeller design	Garmendia design	FDM	ABS/PLA	Blower-based device, 3D printed control panel	Garmendia et al. (2020)
	VentilAid	FDM	ABS/PLA	Impeller-based design, with 3D printed housing	Ventilaid (2020)
	AdamAndEvil	FDM	ABS/PLA	3D printed housing for bespoke bellows	AdamAndEvil (2020)
Solid-state design using compressed gas	ARMEE ventilator	SLA	SLA resin	Solid-state ventilator relying entirely on compressed air to operate without control circuitry or bellows	Ventilator ARMEE (2020)
	Illinois RapidVent	FDM	ABS/PLA	Designed to rely on compressed gas supply to provide ventilation	Branding (2020)
Ventilator flow splitters	Lai design	FDM/SLA	ABS/PLA/SLA resin	Includes inspiratory flow limiter as an attempt to account for differential lung compliance	Lai et al. (2020), Emergency ventilator circuit splitter (2020)
	Clarke design	FDM	ABS/PLA	Two-way splitter	Clarke (2020)
	Chrizz design	FDM	ABS/PLA	Four-way splitter	Chrizz (2020)
	Ventilator Splitter and Resistor System (VSRS)	SLA	SLA resin	Two-way splitter including interchangeable flow resistors with graded lumens	Bishawi et al. (2020)
	VESper ventilator splitter	FDM/SLA	ABS/PLA/SLA resin	Simple two-way splitter	Prisma Health Innovation: VESper (2020), Swathwood (2020)
PEEP valves	PandaPEEP valve	FDM	ABS/PLA	Advanced 3D printed, spring-loaded adjustable PEEP valve	dtextor (2020)
	no2covid valve	FDM	ABS/PLA	External restrictor for existing ventilator tubing	no2covid (2020)

the bellows, with one 3D printed example powered by a swappable drill battery (ThumperX 2020). A number of these 3D printed designs have moved past the prototyping stage to working models ready for testing and production. Interestingly, several designers have scavenged their 3D printers themselves to actuate the BVM (Ad Hoc Ventilator 2020; keeevvviiinnn 2020), although this form of design does not appear to have been subject to any form of experiments with test lungs, animals, or humans.

The Canada-based OpenLung ventilator project (OpenLung Open Source Ventilator 2020) (not to be confused with the open lung ventilation strategy, which combines lung-protective ventilation with high PEEP) provided one of the early forums for discussion of these devices. Their GitHub page coordinates the efforts of various hobbyists and engineers regarding all aspects of the design including mechanics, electronics, and control software (OpenLung 2020). Multiple designs are catalogued including belt systems, rollers, and jaws used to compress the bag. Results of testing are lacking, however, and the OpenLung project remains more of a forum for brainstorming designs rather than a coordinated effort to build a device. The founders of the OpenLung ventilator project have since joined a team of engineers from the Republic of Ireland and the Irish Health Service to attempt to deliver a working model, termed the OSVentilator project (OSV Open Source 2020). This project has received support and donations from several companies. OSVentilator is now reported to have produced several prototype models, and the firmware currently remains under development, although no experimental results are available at the time of writing.

The MIT E-Vent (short for Emergency Ventilator) consists of a BVM suspended between a set of flat paddles, which contract to squeeze the bag and give a breath (MIT Emergency Ventilator Project 2020). The ventilator is borne out of a 2010 project by MIT incorporating a BVM held in a box, squeezed by a cam controlled by an Arduino microcontroller, allowing for setting of tidal volume, respiratory rate, and I:E ratio (Al Husseini et al. 2010). The design also includes a pressure limiting valve, a critical feature of any ventilator. Impressively, the original design featured a mode to support spontaneous breathing with an enforced mandatory minute volume, using pressure triggering to deliver assisted spontaneous breaths. The same pressure sensor doubles as an over-pressure alarm. A mechanical pressure relief valve is stated as a planned adaptation for future versions of the device. The current E-Vent model is reported to have been tested in four porcine experiments, although it should be noted that these experiments are published on the project website, and have not been peer-reviewed. In their experiments, comparison with a Medtronic transport ventilator revealed similar operations and results even in spontaneous mode. Anaesthetists were able to manipulate blood gas parameters of the animal subjects using the ventilator, and even operators not familiar with the design were able to provide ARDS-style ventilation (albeit in animal models with normal lungs).

The Spiro Wave ventilator has been developed as a spin-off from the MIT E-vent, with a view to prepare the design for mass production (Emergency Ventilator Response 2020). The device allows for adjustment of tidal volume, PEEP (up to 25 cm H_2O), respiratory rate, I:E ratio, and airway pressure. The device does feature several safety alarms, such as low and high pressure, high resistance, high driving pressure, incomplete breath, and mechanical failure. One innovative feature is the ability to easily remove the BVM from the device without breaking the circuit. Thus, in the case of mechanical failure or other issues, the BVM can be rapidly detached and used to manually ventilate the patient. It is notable, however, that power failure results in no alarms, and the device does not appear to have an indwelling battery. The design gained FDA emergency use authorisation in April 2020, although their website states that the device is "not FDA cleared or approved". Production has begun in partnership with a number of companies, with the aim to provide ventilators to hospitals in New York City, USA. Their website states that the designers are working with the New York City Economic Development Corporation to distribute over 3000 devices to NYC hospitals, with companies reportedly ready to produce over 500 units per day (Fretty 2020). These statements have not been independently verified, however, and to date there are no public or peer-reviewed reports of the devices being used in clinical practice. While the Spiro Wave project is one of the successes of an originally open-source ventilator device now having undergone commercial partnership and manufacture, it is noteworthy that the project is no longer a true open-source project; it is now a "managed open source" project, providing design resources freely to interested manufacturing parties only. No experimental detail on the device is available, and it is unclear whether distribution or patient testing has occurred at the time of writing.

Other devices are being manufactured by industry, such as the originally open-source VentilatorPal (freebreathing.org 2020) now being produced commercially in the form of the FRD-e by a company known as Stogger based in the Netherlands (Stogger Medical 2020a, b). The VentilatorPal remains open source, with firmware and design files freely available through their website. Despite this, little in the way of specifications of the ventilator are available, and the Wiki page hints towards an inability to sustain the high ventilatory pressures often required in ARDS ventilation, citing the possibility of normal compliance COVID-19 ARDS as a reason that this shortcoming has not been addressed (van der Bij 2020). Stogger provides slightly more technical information on their FRD-e device, describing an optional internal battery, adjustable tidal volumes from 150 to 760 mL, respiratory rate 10–60 breaths per minute, PEEP of 0–20 cm H_2O (set via a breather valve), and maximum pressure of 40 cm H_2O (Stogger Medical 2020b). Despite the somewhat limited technical details and lack of any published tests, the device boasts an impressive and innovative ability to monitor and control multiple ventilators from an Android or iOS app. This may allow for better adherence to infection control procedures, and more stringent awareness of alarms and alerts. The Open Breath project follows a similar principle, but is at an even earlier stage of design (OpenBreath 2020).

Some of the most robust peer-reviewed published testing of a pandemic ventilator comes from Dhanani et al., who managed to successfully ventilate a pig for 12 consecutive hours using the second version of their 3D printed prototype ventilator (Dhanani et al. 2020). The frame was constructed from CNC cut plywood, along with 3D printed gears made from PLA. It is noteworthy that the first version failed twice during the experiment; after 6 h, both PLA and ABS gears failed, primarily due to high stepper motor temperatures during testing with high PEEP resulting in softening of the plastic. The gears were eventually replaced with nylon, and subsequently tested in vitro for 72 consecutive hours. Importantly, the BVM utilised during the second experiment was reported to have shown some signs of wear, a key drawback of all ventilators of this class. Pigs were treated in a crossover study with a standard ventilator and the prototype ventilator, with the authors finding no significant difference between physiological parameters or chest radiographs after ventilation. However, it should be noted that the subjects in the experiments had healthy lungs, and further testing on an ARDS model would be required; the authors propose that their ventilator could be used in a non-COVID-19 patient, such as a patient undergoing elective surgery. Impressively, the device was manufactured in approximately 2 hours, so conceivably devices such as this could be used to facilitate 'just-in-time' supply of medical technology, although a distinct disadvantage of this approach is the lack of rigorous testing for each individual device manufactured.

The Spanish Leitat 1 ventilator is the only BVM-class pandemic ventilator device to date which claims to have been tested on a patient suffering from COVID-19. The machine was reportedly tested at Hospital Parc Taulí, Barcelona, Spain, and has been laboratory tested in maximum settings up to 48 hours (Leitat Managing Technologies 2020). Subsequent to this, it has been approved for use by the Spanish Medical Agency. Their website states that the device has passed animal tests for 6 hours, and has been tested on real patients for two hours (Leitat Managing Technologies 2020). The device was developed in approximately 1 week, is capable of continuously monitoring the machine state, and has a comprehensive set of alarms. An in-built battery powers the alarm systems, and while it does not appear that the battery would be sufficient to ventilate the patient in the event of a power failure, it will alert the operator to the failure of ventilation.

4.2 Non-BVM Designs

Given that an off-the-shelf BVM provides all the necessary valves and pneumatic components required for basic ventilation, it is no wonder that most of the hobby-level and even some commercial designs focus on this approach. However, these devices suffer from a number of serious design issues when used for continuous ventilation. Firstly, it is not known whether the BVM devices can sustain the days of continuous use required of a critical care ventilator. In addition, accurate delivery of reliable and repeatable pressures and tidal volumes may be impossible, and the complex mechanical components required to compress the bag may result in a high failure rate, as has been observed by Dhanani et al. (2020). Given the possible bottleneck of availability of BVM devices in the face of a global crisis, other design ideas are being sought, including utilising motor-driven impellers and compressors (Garmendia et al. 2020), or indeed the use of other components as bellows. While this allows for more flexibility in design, it comes with the disadvantage of a possible increased requirement for medical testing. Thus, while non-BVM devices likely represent the most viable options for real-world operation, their manufacture is fraught with difficulty.

A Polish group called VentilAid in conjunction with the FDM printing company Urbicum has developed an open-source 3D printed ventilator that does not rely on a BVM device, but instead uses an impeller (Ventilaid 2020). The group is now attempting to manufacture the device commercially, and as of August 2020, they are reportedly seeking regulatory approval. However, no specifications or tests have been published. In addition, a technically impressive hobbyist design using bespoke bellows manufactured from vacuum hose exists (AdamAndEvil 2020), but is not tested on humans or animals and does not appear to provide any digital readout of ventilator settings, which are adjusted by liquid weights placed at different positions on the device. It is likely, however, that the real-world clinical applications of this device are limited.

The OpenVentilator Spartan Model (Popsolutions 2020; openventilator 2020) uses a car windscreen wiper motor, and operates without a microcontroller. With this, it has no alarms, and no way of adjusting many of the parameters. An inner tube from a heavy goods vehicle tyre coupled with a slider-crank mechanism acts as a bellows. Some control of PEEP, I:E ratio, and inhale and exhalation time is available. Pressure is controlled by a simple over-pressure valve, and an inhalation bypass valve allows inspiration during the exhale cycle, thus reducing the possibility of barotrauma in a partially awake patient. The machine, however, will not provide any pressure support for a spontaneous breath, and may allow loss of PEEP. This machine also provides no readout regarding patient or ventilator parameters. Pressures are controlled by a manual valve involving a pipe pressed into a bubble chamber, and tidal volume is adjusted by changing the crank arm length, although ventilation must be paused in order to facilitate this. Speed of exhalation and inhalation are adjusted via a diode ladder to control respiratory rate and I:E ratio, using a microswitch to sense the difference between 'inhale' and 'exhale' halves of the main wheel and adjust motor speed accordingly. FiO_2 cannot be controlled. Filtration is achieved by passing exhaust gas through a tank containing bleach, as well as further sterilisation with ultraviolet light. The modular valve block assembly is the main innovative feature of the device; formed of LASER cut flat pieces of silicone and acrylic sandwiched in layers to form check valves, it is reproducible from freely available parts with a LASER cutter, and can be sterilised between patients if de-constructed. The machine has now reportedly been trialled on a test lung simulator. Indeed, the concept of repurposing car parts to construct ventilators is being investigated by the Spanish car company SEAT, which is reportedly repurposing their windscreen wiper motors in their own effort to manufacture a BVM-style pandemic ventilation known as the OxyGEN (Matorell 2020).

Another innovative open-source BVM-free ventilator is known as the Automatic Respiration Management Exclusively for Emergencies (ARMEE) ventilator (Ventilator ARMEE 2020). This is a solid-state device based on a 1965 United States Army design that relies exclusively on compressed gas to operate. The ventilator is calibrated by a number of set screws in order to set pressures, and is designed to oscillate between two states (high and low pressure) in response to the patient's own breathing. Attaching a vacuum to one port can force the machine into the high-pressure state, thus forcing a breath. The project website describes some tested on animals and humans, although no results have yet been made available. The design has the advantage of being very easy to 3D print and maintain, although allows for only minimal control of ventilator settings, no sensing of machine state, no alarms, and is wasteful of oxygen. A similar 3D printed design exists in the form of the Illinois RapidVent, a hand-held ventilator designed to rely only on compressed gas to operate (Branding 2020).

The VITAL Ventilator (Ventilator Intervention Technology Accessible Locally) developed by the NASA Jet Propulsion Laboratory is the pandemic ventilator solution closest to the function of a true ICU ventilator, while still amenable to mass manufacture. The ventilator was developed in just 37 days between March and April 2020, and was tested on 21st March 2020 in the Human Simulation Lab in the Department of Anesthesiology, Mount Sinai Hospital, New York City, USA, reportedly having performed well in these tests (Good 2020). FDA emergency use authorisation was granted on 30 April 2020 (Good et al. 2020a) Following this, NASA stated that 28 companies have been licensed to manufacture the ventilator (NASA Jet Propulsion 2020), including at least one in Brazil reported to have commenced production (Good et al. 2020b). Two versions of the ventilator exist: one with an integrated gas compressor to be used in settings without a medical gas supply, and one for use with compressed gas lines. This is not a true open-source project, however, as the design and even detailed specifications of the device are not available at the time of writing without applying for a free licence. The ventilator attempts to use parts not commonly utilised in the traditional ventilator supply chain in order to avoid competition with existing manufacturers. The designers state that while the VITAL ventilator is designed for use in critical COVID-19 patients, those with the most severe ARDS or more complex respiratory requirements will likely require a fully featured ICU ventilator. However, the available specification is impressive: the machine is capable of FiO_2 of 21–100%, PEEP of 5–35 cm H_2O, respiratory rate of 4–40 breaths per minute, and tidal volume of 150–800 mL. The device features an intuitive panel reporting current settings including FiO_2, PEEP, target tidal volume, peak pressure, respiratory rate and apnoea, inspiratory time, as well as a suite of alarms. The ventilator is reported to have been tested for 20 days of continuous operation without maintenance, and is thought to be usable for 4 months in total before the unit must be replaced, an incredible feat for a pandemic ventilator. Interestingly, the device excludes patient exhalation gas from the circuit to enable the ventilator to be used on multiple patients in sequence without the requirement for sterilisation between patients.

5 Multi-patient Solutions

Given the long lead time required to develop and produce a functional critical care ventilator, some authors have proposed the use of a single ventilator for multiple patients during a surge. There is at least one report of this method being used on humans; a news article briefly mentions multi-patient ventilation in a 2017 mass casualty event in Las Vegas, Nevada, USA, though the article is written in the first person and no peer-reviewed account of this incident is available (Menes et al. 2020). It should also be noted that these trauma victims likely had no underlying lung pathology, which dramatically simplifies the ventilation dynamics. Multiple drawbacks to this approach exist. Differing compliance between patient lungs will inevitably result in differential pressures and volumes. As well as this, ventilating two or more patients at once effaces any useful readout from the ventilator in terms of flow, pressures, and tidal volume (although some attempts have been made to rectify this (Darowski and Englisz 2000; Lee et al. 2020; Clarke et al. 2020)). While there has been some discussion of the relative merits of pressure versus volume control modes for multi-patient ventilation, the overarching issue is that lungs with greater compliance will expand more than lungs with lower compliance at a given pressure, meaning that the patient with greater compliance will be subject to barotrauma, and the patient with lesser compliance will be under-ventilated, with no way of monitoring either (Branson et al. 2012). In addition, the collection of patients ventilated together must be prevented from taking spontaneous breaths by subjecting them to neuromuscular blockade, a treatment usually reserved for those with the most severe and refractory ARDS to facilitate aggressive ventilation modes, and one which is under other circumstances detrimental to recovery. In addition, potential cross-infection issues are associated with this method, especially in the context of COVID-19 respiratory secretions and aerosolisation.

This multi-patient approach has been suggested as early as the 1990s (Sommer et al. 1994), although most published tests to date have relied on healthy subjects (Smith and Brown 2009), or test lungs of similar compliance. One of the early studies used four simulated casualties to test a bespoke ventilator splitter, finding that overall peak pressures were below 35 cm H_2O at all times, and total tidal volume was below 7 mL/kg for all simulated patients (Neyman and Irvin 2006). They found no evidence of differential filling or breath stacking, although it should be noted that there was no accounting for differential compliance in this study, and there was no way to measure the actual volumes and pressures delivered to individual patients. The ventilator splitting approach has been tested in healthy sheep for approximately 12 hours of ventilation (Paladino et al. 2008), although the success of this study given the reported blood gas analysis has been questioned (Branson and Rubinson 2008). Experiments in test lungs have demonstrated that alterations in individual test lung compliance result in wide variation in individual tidal volume (Branson et al. 2012). An updated splitting method by Lai et al. (2020) seeks to resolve the compliance issue by adding an inspiratory flow limiter, the design of which is available online (Emergency ventilator circuit splitter 2020), but once again there is no real-time readout as to whether this is effective. These methods do not allow for precise control of the respiratory parameters of each individual patient, however, although some effort has been made to at least monitor the respiratory parameters of these patients (Englisz and Darowski 2000). Despite the obvious drawbacks, 3D printed versions of ventilator splitters have continued to abound (Clarke 2020; Chrizz 2020), including one innovative solution which includes 3D printed flow restrictors to compensate for lungs of differential compliance (Bishawi et al. 2020). Some further devices for individualising pressures and volumes delivered by a single ventilator include the PANDA peep valve (dtextor 2020), and the no2covid (2020) PEEP restrictor valve.

Although the FDA has apparently issued a 'no objection' statement to the dubious practice of ventilator splitting during the COVID-19 pandemic, and have even authorised use of the VESper ventilator splitter for use during COVID-19 (Prisma Health Innovation: VESper 2020; Swathwood 2020), multiple societies and manufacturers remain adamantly against ventilator splitting in clinical practice due to the issues outlined above (ERCI Clinical Evidence 2020).

6 Conclusion and Future Directions

Makers throughout the globe are harnessing their creativity during worldwide lockdowns in order to attempt solutions to the ongoing COVID-19 pandemic. Many large manufacturing companies are also involving themselves with this effort, either by producing commercial versions of originally open-source designs, or by creating their own de novo ventilator solutions. The difficulty of ARDS ventilation should not, however, be underestimated; ventilatory strategies are still being explored, and patients with COVID-19 ARDS are notoriously difficult to treat. It should be kept in mind that the vast majority of the ventilator prototypes discussed above do not meet even the most basic operational standards set out by the MHRA. Ventilators are extremely complex devices, and unlike other devices that may be manufactured with a failure state in which activity is ceased and the patient comes to no harm, ventilators must operate continuously if the patient is to survive. Even the more robust pandemic ventilators above do not carry the sophisticated settings required to ventilate the complex ARDS patient in ICU.

Pandemic ventilators are often described as 'bridge' ventilators, referring to their intended use when a patient must be ventilated immediately and until an ICU ventilator becomes available. In contrast to type 2 respiratory failure due to failure of the patient's own muscles of respiration, ventilation in type 1 respiratory failure due to damage to alveolar membranes may sometimes be delayed for several hours if clinically appropriate, as self-ventilation on supplemental oxygen is preferable to substandard mechanical ventilation. Early use of an inadequate ventilator may also cause the patient's clinical condition to worsen whilst awaiting an ICU ventilator. Thus, the concept of bridge ventilation is incorrect. Instead, these ventilators should be considered as alternative and step-down ventilators, in which context they may be used as alternatives for simple surgical anaesthetics in which patients have normal lungs, for instance, freeing more complex anaesthetic ventilators for patients in ICU. In patients whose acute COVID-19 ARDS has resolved, they will often be subject to prolonged weaning from ventilatory support in order to build respiratory muscle bulk and encourage self-ventilation. Patients commonly undergo tracheostomy weaning programmes, including daily periods off ventilation, with only pressure support from a ventilator overnight. These patients would be ideal candidates for a pandemic ventilator that has less functionality than an ICU ventilator, and are potentially more prone to failure, as the patients are capable of self-ventilating, and thus ventilator failure may be more tolerable.

There are many further avenues for exploration of critical care technology for use during a pandemic surge. Each ventilated patient with ARDS will require multiple infusions of medications and nutrition; these pumps and consumables are comparatively simple in design, although absolutely vital for critical care, and thus represent an attractive but under-explored target for makers and hobbyists. Haemofiltration devices rely on closed circuits designed by mainstream manufacturers and available in scale, whereas the comparatively simple machinery which pumps blood through the circuits is in short supply. Thus, the open-source community is primed for developing tools for use in fighting COVID-19, with several devices already at advanced stages of development. Rather than focus on one of the most complex and critical pieces of medical technology, however, it would be useful for makers to turn to other simple critical care supply issues that may have a profound impact on the care of patients throughout the COVID-19 pandemic and beyond.

References

Ad Hoc Ventilator (2020) Ad Hoc ventilator. https://20100.be/ad-hoc-ventilator/

AdamAndEvil (2020) Ventilator prototype. Thingiverse database, MakerBot Industries, Brooklyn, NY, USA. https://www.thingiverse.com/thing:4538770

Al Husseini AM, Lee HJ, Negrete J, Powelson S, Servi AT, Slocum AH et al (2010) Design and prototyping of a low-cost portable mechanical ventilator. J Med Devices 4(2):027514

Alsema A (2020) Colombia close to having world's first open source and low-cost ventilator to 'beat Covid-19'. Colombia Reports, Colombia; Carlota V (2020) Medically approved emergency 3D printed ventilator goes into production. 3Dnatives. https://colombiareports.com/colombia-close-to-having-worlds-first-open-source-and-low-cost-ventilator-to-beat-covid-19/

AmboVent (2020) The low-cost, build-it-anywhere ventilator. https://ambovent.org

Bishawi M, Kaplan M, Chidyagwai S, Cappiello J, Cherry A, MacLeod D, et al (2020) Rapid ventilator splitting during COVID-19 pandemic using 3D printed devices and numerical modeling of 200 million patient specific air flow scenarios. Res Sq

Branding T (2020) Army researchers help design 3D-printed ventilators. Army News Service, USA. https://www.army.mil/article/234822/army_researchers_help_design_3d_printed_ventilators

Branson RD, Rubinson L (2008) One ventilator multiple patients—what the data really supports. Resuscitation 79(1):171–172

Branson RD, Blakeman TC, Robinson BR, Johannigman JA (2012) Use of a single ventilator to support 4 patients: laboratory evaluation of a limited concept. Respir Care 57(3):399–403

Brower RG, Matthay MA, Morris A, Schoenfeld D, Thompson BT, et al (2000) Ventilation with lower tidal volumes as compared with traditional tidal volumes for acute lung injury and the acute respiratory distress syndrome (Acute Respiratory Distress Syndrome Network). N Engl J Med 342(18):1301–1308

Carlota V (2020) Medically approved emergency 3D printed ventilator goes into production. 3Dnatives. https://www.3dnatives.com/en/3d-printed-respirator-230320205/

Chandler DL (2020) In the world: breath of life. https://news.mit.edu/2010/itw-ventilator-0715

Chong MSF et al (2020) ICU surge capacity in a busy London district general hospital during the COVID-19 pandemic. Association of Anaesthetists, London, United Kingdom. https://anaesthetists.org/Home/Resources-publications/COVID-19-guidance/ICU-surge-capacity-in-a-busy-London-district-general-hospital-during-the-COVID-19-pandemic
Chrizz (2020) Breathing ventilator H-splitter. Thingiverse database, MakerBot Industries, Brooklyn, NY, USA. https://www.thingiverse.com/thing:4235285
Clarke AL (2020) 3D printed circuit splitter and flow restriction devices for multiple patient lung ventilation using one anaesthesia workstation or ventilator. Anaesthesia 75(6):819–820
Clarke AL, Stephens AF, Liao S, Byrne TJ, Gregory SD (2020) Coping with COVID-19: ventilator splitting with differential driving pressures using standard hospital equipment. Anaesthesia 75(7):872–880
Costa D (2020) Mechanic ventilator mini—COVID 19. Thingiverse Database, MakerBot Industries, Brooklyn, NY, USA. https://www.thingiverse.com/thing:4245897
Covid19-vent-list (2020) GitHub, San Francisco, CA, USA. https://github.com/PubInv/covid19-vent-list
Darowski M, Englisz M (2000) Artificial ventilation of the lungs for emergencies. Front Med Biol Eng 10(3):177–183
deBoisblanc BP (2005) Black Hawk, please come down: reflections on a hospital's struggle to survive in the wake of Hurricane Katrina. Am J Respir Crit Care Med 172(10):1239–1240
Dhanani J, Pang G, Pincus J, Ahern B, Goodwin W, Cowling N et al (2020) Increasing ventilator surge capacity in COVID 19 pandemic: design, manufacture and in vitro-in vivo testing in anaesthetized healthy pigs of a rapid prototyped mechanical ventilator. BMC Res Notes 13(1):421
dtextor (2020) PANDApeep Gen3 PEEP valve. Thingiverse database, MakerBot Industries, Brooklyn, NY, USA. https://www.thingiverse.com/thing:4278384
Duggal A, Pinto R, Rubenfeld G, Fowler RA (2016) Global variability in reported mortality for critical illness during the 2009–10 influenza A(H1N1) pandemic: a systematic review and meta-regression to guide reporting of outcomes during disease outbreaks. PLoS ONE 11(5):e0155044
Emergency ventilator circuit splitter (2020) ventsplitter.org. https://ventsplitter.org
Emergency Ventilator Response (2020) We are responding to the COVID-19 ventilator shortage by rapidly scaling the production and distribution of Spiro Wave. https://ventilatorresponse.com
Englisz M, Darowski M (2000) Artificial ventilation of the lungs for emergencies. Frontiers Med Biol Eng 10(3):177–183
ERCI Clinical Evidence Assessment (2020) Single Ventilator use to support multiple patients. Elsevier, Amsterdam, Netherlands. https://www.elsevier.com/__data/assets/pdf_file/0009/1000314/COVID-ECRI_HTA_Single-Ventilator-Use-Multiple-Patients.pdf
Force ADT, Ranieri VM, Rubenfeld GD, Thompson BT, Ferguson ND, Caldwell E, et al (2012) Acute respiratory distress syndrome: the Berlin definition. JAMA. 307(23):2526–2533
Frazer JS, Tyrynis Everden AJ (2020) Emerging patterns of hypercoagulability associated with critical COVID-19: a review. Trends Anaesth Crit Care S2210844020300903
freebreathing.org (2020) VentilatorPAL. FreeBreathing.org. https://freebreathing.org
Fretty P (2020) What happens when a crucial supply chain fails?. Industry Week, Endeavor Business Media, Nashville, TN, USA. https://www.industryweek.com/covid19/article/21129254/what-happens-when-a-crucial-supply-chain-fails
Garmendia O, Rodríguez-Lazaro MA, Otero J, Phan P, Stoyanova A, Dinh-Xuan AT, et al (2020) Low-cost, easy-to-build noninvasive pressure support ventilator for under-resourced regions: open source hardware description, performance and feasibility testing. Eur Respir J 55(6)
GlobalData (2020) 880,000 global ventilator shortage can only be met by new simple design requiring minimal operator training time, says GlobalData. https://www.globaldata.com/880000-global-ventilator-shortage-can-only-be-met-by-new-simple-design-requiring-minimal-operator-training-time-says-globaldata/
Good A (2020) NASA develops COVID-19 prototype ventilator in 37 days. NASA, Washington, D.C., USA. https://www.nasa.gov/feature/jpl/nasa-develops-covid-19-prototype-ventilator-in-37-days

Good A et al (2020a) NASA-developed ventilator authorized by FDA for emergency use. NASA, Washington, D.C., USA. https://www.nasa.gov/press-release/nasa-developed-ventilator-authorized-by-fda-for-emergency-use

Good A et al (2020b) Brazilian partnership to begin producing NASA-designed COVID-19 ventilator. NASA, Washington, D.C., USA. https://www.nasa.gov/feature/jpl/brazilian-partnership-to-begin-producing-nasa-designed-covid-19-ventilator

Gruber PC, Gomersall CD, Joynt GM (2006) Avian influenza (H5N1): implications for intensive care. Intensive Care Med 32(6):823–829

grumpytacos (2020) MIT E-Vent v2—printable with HybridStepper. Thingiverse database, MakerBot Industries, Brooklyn, NY, USA. https://www.thingiverse.com/thing:4390796

Hewing L, Menner M, Tachatos N, Daners MS, Pasquier C du, Lumpe TS, et al (2020) Volume control of low-cost ventilator with automatic set-point adaptation. http://arxiv.org/abs/200901530

joelfrax (2020) Frax3D ventilator 2.0—Ambu. Thingiverse database, MakerBot Industries, Brooklyn, NY, USA. https://www.thingiverse.com/thing:4255247

keeevvviiinnn (2020) 3DPaV (3D-Printer-as-Ventilator) parts [Non-Human Use Only]. Thingiverse database, MakerBot Industries, Brooklyn, NY, USA. https://www.thingiverse.com/thing:4566892

Kleinman Z (2020) Coronavirus: 3D printers save hospital with valves. BBC, London, England, UK. https://www.bbc.co.uk/news/technology-51911070

Lai BK, Erian JL, Pew SH, Eckmann MS (2020) Emergency open-source three-dimensional printable ventilator circuit splitter and flow regulator during the COVID-19 pandemic. Anesthesiology 133(1):246–248

Lassen HCA (1953) A preliminary report on the 1952 epidemic of poliomyelitis in Copenhagen with special reference to the treatment of acute respiratory insufficiency. Lancet 1(6749):37–41

Lee A, Umar S, Hoftman N (2020) A novel multi-ventilation technique to split ventilators. Anesthesia. http://medrxiv.org/lookup/doi/10.1101/2020.04.28.20083741

Leitat Managing Technologies (2020a) The first medically validated and industrially producible 3D emergency ventilator. Leitat Managing Technologies. https://covid-leitat.org/en/ventilator/

Leitat Managing Technologies (2020b) Frequently asked questions about LEITAT 1 ventilator. Leitat Managing Technologies. https://covid-leitat.org/en/faqs-leitat-1-3/

Matorell (2020) From making cars to ventilators. SEAT, Spain. https://www.seat-mediacenter.com/newspage/allnews/company/2020/From-making-cars-to-ventilators.html

Matthay MA, Zemans RL, Zimmerman GA, Arabi YM, Beitler JR, Mercat A, et al (2019) Acute respiratory distress syndrome. Nat Rev Dis Primers 5(1):18

Medicines and Healthcare products Regulatory Agency (2020) Rapidly manufactured ventilator system. Medicines and Healthcare products Regulatory Agency, London, United Kingdom. https://assets.publishing.service.gov.uk/government/uploads/system/uploads/attachment_data/file/879382/RMVS001_v4.pdf

Menes K, Tintinalli J, Plaster L (2020) Emergency physicians monthly. EPMonthly.com, USA. https://epmonthly.com/article/not-heroes-wear-capes-one-las-vegas-ed-saved-hundreds-lives-worst-mass-shooting-u-s-history/

MIT (2020) Key ventilator specifications. Massachusetts Institute of Technology, Cambridge, MA, USA. https://e-vent.mit.edu/clinical/key-ventilation-specifications/

MIT Emergency Ventilator Project (2020) MIT Emergency Ventilator, MIT, MA, USA. https://emergency-vent.mit.edu

Monolithic Power Systems (2020) MPS open-source ventilator. Monolithic Power Systems, San Jose, CA, United States. https://www.monolithicpower.com/en/mps-open-source-ventilator

NASA Jet Propulsion Laboratory (2020) VITAL—the COVID-19 ventilator device. NASA, Washington, D.C., USA. https://medeng.jpl.nasa.gov/covid-19/ventilator/registration/

Neyman G, Irvin CB (2006) A single ventilator for multiple simulated patients to meet disaster surge. Acad Emerg Med 13(11):1246–1249

no2covid (2020) no2covid. https://no2covid.com

OpenBreath (2020) OpenBreath. https://www.openbreath.it/en/

OpenBuilds (2020) V-Slot ventilator. Thingiverse database, MakerBot Industries, Brooklyn, NY, USA. https://www.thingiverse.com/thing:4301467

OpenLung (2020) GitHub, San Francisco, CA, USA. https://gitlab.com/open-source-ventilator

OpenLung Open Source Ventilator (2020) OpenLung. https://openlung.org

OpenVent (2020) OpenVent—3D printable BVM ventilator, design A2, our favorite. Thingiverse Database, MakerBot Industries, Brooklyn, NY, USA. https://www.thingiverse.com/thing:4274786

openventilator (2020) GitHub, San Francisco, CA, USA. https://github.com/popsolutions/openventilator

OSV Open Source Ventilator (2020) OSV, Dublin, Ireland. https://opensourceventilator.ie

Paladino L, Silverberg M, Charchaflieh JG, Eason JK, Wright BJ, Palamidessi N et al (2008) Increasing ventilator surge capacity in disasters: ventilation of four adult-human-sized sheep on a single ventilator with a modified circuit. Resuscitation 77(1):121–126

Pandemic Ventilator Project (2020) Pandemic ventilator project list of all posts. Pandemic Ventilator Project. https://panvent.blogspot.com/2007/02/pandemic-ventilator-project-list-of-all.html

Pearce JM (2020) A review of open source ventilators for COVID-19 and future pandemics. F1000Res 9:218

Petsiuk A, Tanikella NG, Dertinger S, Pringle A, Oberloier S, Pearce JM (2020) Partially reprapable automated open source bag valve mask-based ventilator. HardwareX 8:e00131

Popsolutions (2020) Open ventilator Popsolutions, Brazil. https://popsolutions.co/en_US/openventilator

Prisma Health Innovation: VESper (2020) Prisma Health, South Carolina, USA. https://www.prismahealth.org/VESper/

ProgressTH (2020) Low-cost opensource mechanical ventilator. Thingiverse database, MakerBot Industries, Brooklyn, NY, USA. https://www.thingiverse.com/thing:4292941

Read RL (2020) Evaluating open source ventilator projects. https://docs.google.com/document/u/1/d/e/2PACX-1vRl9yZ27KvslftcNvweHgH1A81pO8gHL62TWpY_VY-UELWdK9x-4-3hNw3DbkemClzExPsg8RfnxilP/pub

Rice University (2020) ApolloBVM—Emergency use ventilator. Rice University, Houston, TX, USA. http://oedk.rice.edu/apollobvm

Ronco C, Reis T, Husain-Syed F (2020) Management of acute kidney injury in patients with COVID-19. Lancet Respir Med 8(7):738–742

Shaver CM, Bastarache JA (2014) Clinical and biological heterogeneity in acute respiratory distress syndrome: direct versus indirect lung injury. Clin Chest Med 35(4):639–653

Sinha P, Calfee CS, Cherian S, Brealey D, Cutler S, King C, et al (2020) Prevalence of phenotypes of acute respiratory distress syndrome in critically ill patients with COVID-19: a prospective observational study. Lancet Respir Med S2213260020303660

Siripanthong B, Nazarian S, Muser D, Deo R, Santangeli P, Khanji MY et al (2020) Recognizing COVID-19-related myocarditis: the possible pathophysiology and proposed guideline for diagnosis and management. Heart Rhythm 17(9):1463–1471

Smith R, Brown JM (2009) Simultaneous ventilation of two healthy subjects with a single ventilator. Resuscitation 80(9):1087

Sommer DD, Fisher JA, Ramcharan V, Marshall S, Vidic DM (1994) Improvised automatic lung ventilation for unanticipated emergencies. Crit Care Med 22(4):705–709

Stogger Medical (2020a) Introducing FRD-e. Stogger Medical, Eindhoven, Netherlands. https://medical.stogger.com

Stogger Medical (2020b) FRD-e ventilation assistant. Stogger Medical, Eindhoven, Netherlands. https://medical.stogger.com/wp-content/uploads/2020/06/Stogger-Medical-FRD-e-Technical-Datasheet.pdf

Swathwood S (2020) FDA approves 'Y' device to split one ventilator between multiple patients. 23 News Week, Illinois, USA. https://week.com/2020/03/31/fda-approves-y-device-to-split-one-ventilator-between-multiple-patients/

ThumperX (2020) Breathing ventilator. Thingiverse database, MakerBot Industries, Brooklyn, NY, USA. https://www.thingiverse.com/thing:4270370

Truog RD, Mitchell C, Daley GQ (2020) The toughest triage—allocating ventilators in a pandemic. N Engl J Med 382(21):1973–1975

University of Minnesota Medical School (2020) A ventilator system built for rapid development. University of Minnesota Medical School, Minneapolis, MN, USA. https://med.umn.edu/covid19Ventilator

van der Bij E (2020) Frequently asked questions about the VentilatorPAL. Open Hardware Repository. https://ohwr.org/project/ventilatorpal/wikis/FAQ

Ventilaid (2020) Ventilaid august update. Ventilaid, Poland. https://www.ventilaid.org/ventilaid-august-update/

Ventilator ARMEE (2020) Automatic respiratory management exclusively for emergencies. ARMEEVent. armeevent.com

Yao W, Wang T, Jiang B, Gao F, Wang L, Zheng H et al (2020) Emergency tracheal intubation in 202 patients with COVID-19 in Wuhan, China: lessons learnt and international expert recommendations. Br J Anaesth 125(1):e28–e37

zzzrrrzz (2020) Emergency ventilator (EV-02). Thingiverse database, MakerBot Industries, Brooklyn, NY, USA. https://www.thingiverse.com/thing:4302479

Artificial Intelligence in 3D Printing: A Revolution in Health Care

Aishwarya Banerjee, Haritha K. Haridas, Arunima SenGupta, and Neetu Jabalia

Abstract In the era of digital world, due to constantly evolving medical technology, Artificial Intelligence (AI) propelled out of research labs into our everyday lives. Emerging developments in AI have an advantage in healthcare system as they will reduce the manual process performed by humans. Hence, AI is a game-changer in health care because when AI is combined with three-dimensional (3D) printing technologies it could increase the performance by reducing the risk of error and facilitating automated production. Nowadays, 3D printing has become an essential and potentially transformative approach to revolutionize health care rapidly. One of the objectives of this chapter is to present an overview of basic concepts of AI including machine learning, internet of things, cloud computing, and deep learning. The further dimensions will also describe types of 3D printing. A snapshot of stereolithography (SLA), which is the oldest and reliable approach in 3D printing is also described. The present chapter will also highlight the key healthcare trends impacting the 3D printing industry such as use of 3D printing to fight the pandemic outbreak COVID-19; regenerative medicine; eliminating the prerequisite for animal experimentation to confirm their efficacy and safety before human testing begins and personalized medicines. The last section encompasses key applications, challenges, and future of 3D printing, which will help the readers to understand that this emerging technology is becoming society's most helpful tool and further may effectively use the information for their research endeavors.

Keywords Healthcare · Artificial intelligence · 3D printing · COVID-19 · Stereolithography · Machine learning

---

A. Banerjee · H. K. Haridas · A. SenGupta · N. Jabalia (✉)
Amity Institute of Biotechnology, Amity University Uttar Pradesh, Noida 201301, India
e-mail: njabalia@amity.edu

K. Sandhu et al. (eds.), *Emerging Applications of 3D Printing During CoVID 19 Pandemic*, Lecture Notes in Bioengineering, https://doi.org/10.1007/978-981-33-6703-6_4

1 Introduction

Technology has led to the digitalization of the healthcare sector with artificial intelligence (AI) enabled medical devices, telemedicine, electronic health records being some of its novel innovations (Fig. 1). It is mainly built on the goal of improving patient results, cost reduction, optimizing systems, and reducing human errors. While on one hand, the doctors are able to store and utilize their resources in a better way using AI-powered systems, patients, on the other hand, are able to get better treatment with wearable medical devices, 5G mobile technology, and virtual reality tools (Reddy and Ram 2020). 3D printing or Additive Manufacturing (AM) enables the manufacturing of patient-specific and custom-designed parts. It can be used to manufacture medical devices for surgeries such as anatomical models, surgical instruments, and orthoses (Chen et al. 2016) with such a high level of complexity and in such a cost-effective manner that will never be possible with the conventional manufacturing methods (Kang et al. 2016). As a result, 3D printing is becoming a widely used model of manufacturing in fields such as tissue engineering, medical devices, dentistry, regenerative medicine, and drug formulation. It is also serving as a popular platform for developing research areas such as tissue and organ printing. Advanced industries use various information technology such as 3D printing, artificial intelligence, cloud computing, machine learning, big data, internet of things, etc. (Lee et al. 2018). Artificial intelligence refers to a PC framework that performs all the functions that usually require human insight. Machine learning (ML) is the subset of artificial intelligence which studies computer algorithm that improves the output. ML makes predictions or decisions with the help of training data or sample data. Therefore, this chapter discusses the overview of 3D printing approaches which are entwined with techniques such as AI, ML, deep learning, cloud computing, and Internet of Things (IoT). The next section will focus on the introduction to 3 D printing, its types, and significant role stereolithography (SLA) which is the oldest

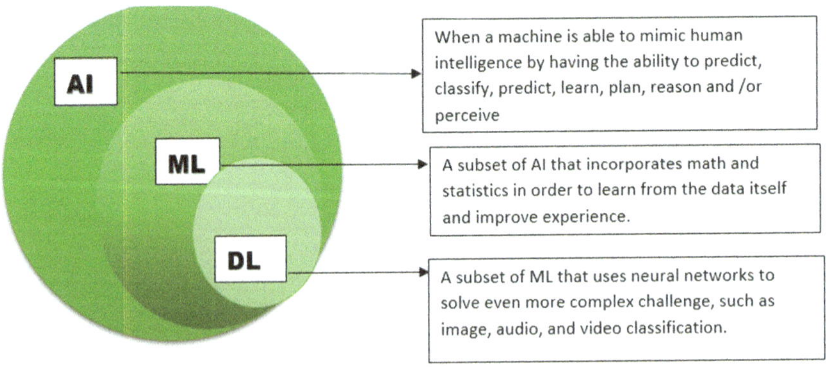

Fig. 1 The relationship between artificial intelligence, machine learning, and deep learning

and reliable approach in AM. It will also highlight the implementation of technologies in the current pandemic outbreak, i.e., COVID-19. Further, this chapter will present the regenerative medicine; eliminating the prerequisite for animal experimentation to confirm their efficacy and safety before human testing begins and personalized medicines along with key applications of AM. The last section emphasizes the challenges and potential of 3D printing which **will empower** the readers to understand that this emerging technology is becoming society's most helpful tool in the healthcare system.

2 Artificial Intelligence

The term Artificial Intelligence (AI) was coined in the year 1950 by John McCarthy. Artificial intelligence is a PC framework that performs all the functions that usually require human insight. Artificial intelligence works using different sets of algorithms. Numerous AI calculations are able to learn from data; they can upgrade themselves by learning new strategies. The various research areas of AI consist of knowledge, representation, planning, learning, reasoning, language processing, and perception (Norvig 2003). A common AI studies various conditions and performs activities that expand its chances of achievement (du Boulay 2001).

2.1 Artificial Intelligence Can Be Mainly Categorized into Two Parts

(a) Narrow AI

It is also known as weak AI. This type of AI works with a limited amount of data and is the replication of human intelligence. Narrow AI is capable of performing a single work efficiently. Although these devices seem to be intelligent, it has many limitations and restrictions. Few examples of narrow AI are, i.e., Google search, Siri, Alexa, Self-driving car, and Image recognition software

(b) Artificial general intelligence (AGI)

This is also known as strong AI. This type of device is capable of solving any problem with general intelligence just as human intelligence. This type of intelligence can be seen in movies like Star Trek, robots, etc. General intelligence refers to the capability of acquiring various goals, performing various tasks in a variety of contexts. AGI has a vast generalization capability (Ben et al. 2014).

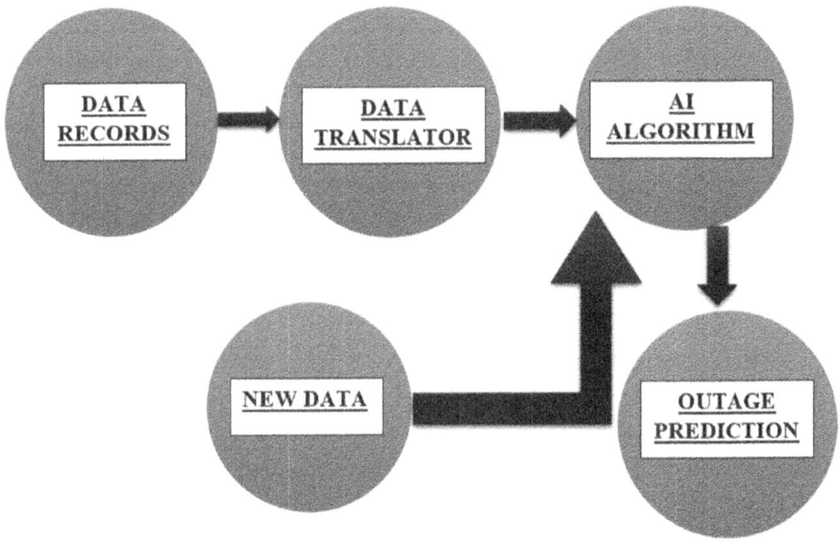

Fig. 2 Working of AI

3 Working of AI

AI program works as first it takes input, then processes it, and gives the required output (Fig. 2). Input is taken from various devices like Sensors, Cameras, etc., an example of the sensor can be Audio Sensors, IR, etc. The process of taking data, all information is gathered in an information base, and the database near the unit starts (compares contribution to working information), since every device is programmed for a particular purpose. If input information coordinates the working information it accomplishes its work by transferring signals to the output. The output part works according to the instructions provided by the processing part. For example, in fingerprint recognition, it will open the computer only when the fingerprints match.

4 Machine Learning

The term machine learning was given by Arthur Samuel. Machine learning is the subset of artificial intelligence which studies computer algorithm, that improves the output. Machine learning makes predictions or decisions with the help of training data or sample data (LeCun et al. 2015). Machine learning is used in various fields such as email filtering, computer vision, etc. we can also say that machine learning is capable of detecting patterns in data, then with the help of uncovered patterns it performs other tasks such as decision-making or detecting patterns in data (Bruggeman 2012).

There are majorly two types of machine learning.

4.1 Supervised Learning

It is also known as "predictive learning approach". If the algorithm of a data set is given with known outputs then it is known as supervised learning. There are various tasks in machine learning, which can be denoted or assigned as supervised. Supervised learning is used in various fields such as bioinformatics, pattern recognition, handwriting recognition, speech recognition, spam detection, etc.

4.2 Unsupervised Learning

It is also known as knowledge discovery. It is more widely used than supervised learning as it does not need any kind of human intelligence to label the data (Heaton 2017). It performs more complex tasks than supervised learning.

5 Working of Machine Learning

The process begins with inserting training data into a given set of algorithms. The training data might be known or obscure data to build up the last algorithm of machine learning. The type of training data used impacts the calculation. To check whether the calculation operates correctly or not, a new set of input data is provided in the machine learning algorithm. Finally, the prediction and results are checked. If the obtained prediction is not as required, then the algorithm is reprocessed several times until the required output is obtained. This helps the machine learning algorithm to learn and develop the most accurate answers and thus helps in increasing the accuracy (Fig. 3).

6 Deep Learning

It is a type of machine learning that allows the computer to learn from previous experiences and understand the data in the form of a hierarchy of logics. Since the PC acquires knowledge from various experiences, therefore there is no requirement of human intervention (Bengio et al. 2013). Deep learning can be further divided into supervised, unsupervised, and semi-supervised learnings (Xia et al. 2012). Deep learning is used in various areas such as visual object recognition, speech recognition, object detection, drug discovery, and genomics (Bak 2003). The word deep in deep learning refers to the presence of several layers in the network. Deep learning uses neural network to copy human intelligence. It can contain up to 150 neural networks.

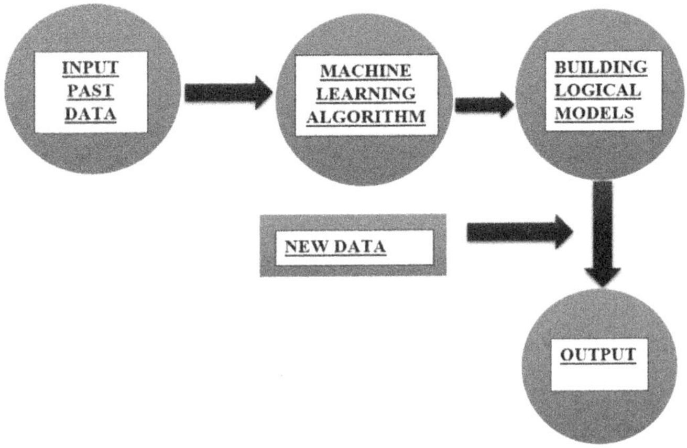

Fig. 3 Working of machine learning

The neural network comprises three parts: input layer, hidden layers, and output layer. CNN is often known as a convolutional neural network.

7 Internet of Things (IoT)

Internet of Things (IoT) refers to the initiation of complex network of items and objects connected through the Internet, often quipped with abundant intelligence (Lakafosis et al. 2010). The following is the connection between IoT and 3D printing (Fig. 4). In stage one, products that are manufactured through 3D printing should have a unique identity code within the object, for example, RFID (Radio-Frequency Identification) tags which are used to identify IoT products (Pandey et al. 2013). Through 3D printing, such unique identification codes can be embedded within the product during the manufacturing process (Löffler et al. 2011). Since RFID technology is very cost-effective, more and more companies are adopting this technique leading to its widespread usage. In stage two, products are embedded with active sensors that are connected by a pre-embedded code. This code makes the product uniquely identifiable and equipped with active sensors. If readable codes can be directly embedded into products, then it means that the object created automatically becomes a part of IoT (Xu 2012). Moreover, since the 3D printers can be connected to the Internet, data can be shared with the environment. Stage three is the natural evolution of stages one and two. Products and 3D printers are both connected to the Internet, which basically means that they lead to a constant flow of data in two ways. The first refers to the product and the second refers to the 3D printers, which can be monitored and controlled remotely as it is connected to the Internet. This concept is very important in terms of smart manufacturing for the development of 'smart factory' (Erol et al. 2016).

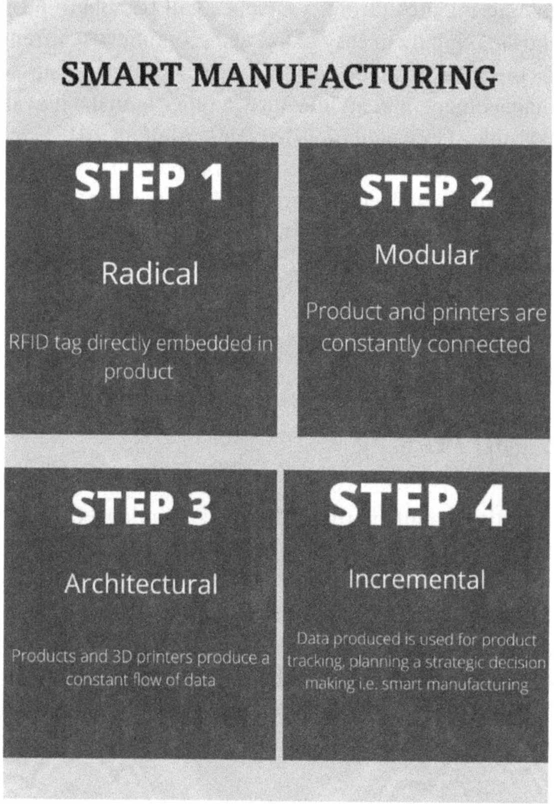

Fig. 4 Steps of smart manufacturing

Therefore, 3D printing can be called 'smart manufacturing' in reference to the technological shift that unites at stage 4. Production can be forecasted and adjusted according to the data provided by the items in terms of demand and specifications (Wu et al. 2013). This process results in 'smart manufacturing' beginning from IoT to 3D printing.

8 Cloud Computing

Cloud manufacturing is a concept that is considered as one of the main drivers of the Industry 4.0 paradigm (He and Xu 2014). Cloud-based manufacturing is basically a service-oriented manufacturing prototype with characteristics such as scalability, customer orientation, and distributive collaboration (Tao et al. 2015). It involves the mapping of manufacturing capabilities and resources into the cloud in the form of services along with providing management and control capabilities to manage the processes, resources, transactions, and operations (Ghobakhloo 2018). It is expected that all manufacturing companies will be able to propose their resources on cloud, thus making them easy to use and economical. Currently, a lot of work is being done in the field of cloud manufacturing in terms of its technologies, concept, service management, and architecture in order to make it available in the market as soon as possible (Ghobakhloo 2018; Alomari et al. 2015). In this context, 3D printing is a very well-suited production method that can be integrated with cloud manufacturing. 3D printing nowadays provides the freedom to produce actual products and this knowledge can be used in cloud manufacturing. Additive manufacturing with its manufacturing design freedom, rapid prototyping, speed of production, and supply chain reductions is expected to help cloud manufacturing reach its maximum capacity (Wang et al. 2017).

9 3D Printing

Additive manufacturing (AM) or 3D printing technology (3DP), also known as digital fabrication or additive manufacturing, refers to the technology of creating physical objects from a geometrical representation through successive addition of materials. In this technology, an object is printed through the layer-by-layer deposition of the raw material through a computer-aided design (CAD) model. It is a fast-growing technology and is currently being used in a lot of industries such as agriculture, locomotive industry, healthcare industry, automotive industry, aviation industry, etc. Fig. 5 represents the working of 3D printing.

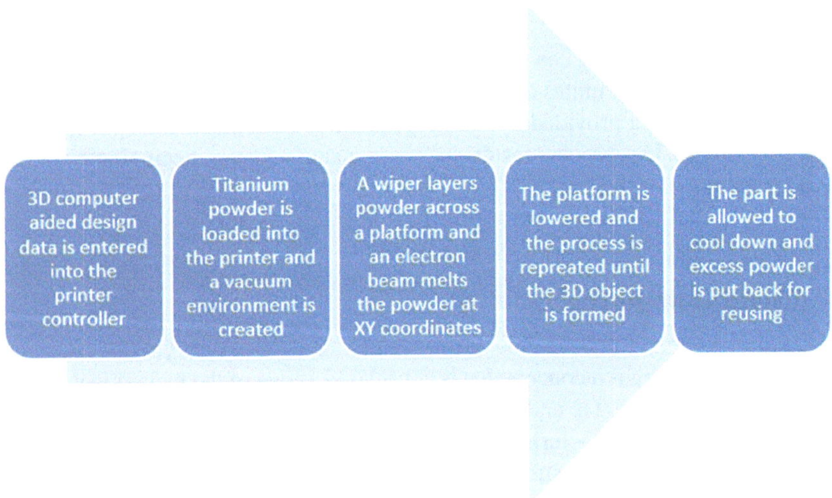

Fig. 5 Systematic representation of 3D printing

3D printing technology holds the potential to revolutionize industries. It can easily increase the speed of production while reducing costs. Consumers can personalize the final product by giving their inputs. Also, since the 3D printing facilities will be located near the consumer, the manufacturing process will be more flexible and responsive. Lastly, 3D printing technologies will change the company logistics as the company can itself manage the full process and can offer better comprehensive and start to finish services (Low et al. 2017).

9.1 Types of 3D Printing

In order to perform a variety of functions, a large variety of technologies related to 3D printing have emerged. According to the American Society for Testing and Materials (ASTM) Standard catalog, there are seven types of AM or 3D printing technologies, viz., vat photopolymerization, binding jetting, sheet lamination, directed energy deposition, power bed fusion, material extrusion, and material jetting (Table 1). Since

Table 1 Types of 3D printing and its description

S.No.	Types of 3D printing	Description
1.	Binder jetting	In binder jetting, layer by layer is formed on top of the spread powder through a jet chemical binder (Shen and Naguib 2019)
2.	Material extrusion	In material extrusion, the products are built starting from the bottom and growing toward the top through heating and extrusion of a thermoplastic filament (Stansbury and Idacavage 2016; Peyre et al. 2017)
3.	Directed energy deposition	Directed energy deposition works in the same way as material extrusion except for that in this process the nozzle is not fixed at a specific axis, but is movable in different directions (Tiwari et al. 2015; Smoczok et al. 2020)
4.	Material jetting	In material jetting, the build material is selectively deposited drop by drop. A print head distributes drops of a photosensitive material which solidifies producing the product layer after layer under ultraviolet light (Tofail et al. 2018)
5.	Power bed fusion	In power bed fusion, either a laser or an electron beam is used to fuse or melt the materials together (Popov et al. 2017; Mau et al. 2019)
6.	Sheet lamination	In sheet lamination technology, several sheets of the raw materials are bonded together in order to produce the final product (Tofail et al. 2018; Livingston et al. 2020)
7.	Vat photopolymerization	In vat photopolymerization, photo-reactive polymers are curated using a UV or a laser light (Tiwari et al. 2015; Mau et al. 2019)

each of these technologies has their individual targeted applications there is no need to debate over which technology or machine is better. Earlier 3D printing technologies were only utilized for prototyping but nowadays it is used mainly for the production of a wide variety of products (Yap et al. 2017).

9.1.1 Stereolithography (SLA)

3D printing is a popular technology which has the ability to construct the structure of accurate geometry and is widely used in the field of tissue engineering. Stereolithography is a very versatile method. Drugs can be combined with photopolymer before printing and can become a hardened matrix. In comparison to other AM methods, SLA is way more superior in relation to resolution (Arcaute et al. 2010). In SLA, there is minimal localized heating during the process of printing which makes it more feasible to use. The drawback of SLA technology is that the photocrosslinkable polymer is limited in number also these materials are not recognized as safe (Kalim and Syed 2020). In this methodology, the 'printhead', laser beam is centered into a tank of fluid resin. This laser results in photopolymerization of the liquid resin, framing a cross-connected polymeric grid and thus the creation of a strong mass. Once more, objects are created by hardening them in a layer-by-layer measure (Arcaute et al. 2010). SLA-AM has been broadly utilized in the field of tissue designing (Tibbits 2016). One of the huge and major advantages of using SLA printing is that the dynamic fixing as well as any excipients can be consolidated into the fluid as long as they are mixed easily. It doesn't make a difference whether the medication or excipients have photo-polymerizable as their functional group; any extra parts basically become caught in the polymeric network. Polymers incline to respond to the use of an external agent and form a group of substances that are suitable for various applications in different research areas. The pH-responsive polymers have a great scope in 3-dimensional and 4-dimensional printing. 3D and 4D printing uses various polymers that include pH-responsive polymers (Bateman and Leach 1998).

10 Use of 3D Printing Technology to Fight the Outbreak of COVID-19

With the recent outbreak of COVID-19, a key shortage of equipment like hoods, masks, and PPE kits have been reported by hospitals. Moreover, the fear of this disease has caused people to hoard necessary items for themselves, leaving others such as health workers in a limited supply. By the end of January 2020, World Health Organization (WHO) declared it as an international "public-health emergency" as the number of COVID-19 patients kept increasing with each passing day (Rochman 2020). In order to curb the problem of shortage of protective kits, several

3D printing communities along with engineers and physicians have started manufacturing a variety of reusable personal protective equipment devices for public supply. AM provides a cost-effective and rapid method of manufacturing lightweight plastic frameworks based on open-source data. Venture valves, a key component of the respiratory support system (Jurischka et al. 2020), is very difficult to substitute or replace at such short notice. Face masks used to protect the person wearing it from airborne viruses are also facing a shortage of supply. The US FDA has issued a 'no objection' policy in regard to manufacturing of medical kits for COVID-19, which for now is applicable throughout the pandemic situation. As a result, several companies, laboratories, individual users have successfully made face masks using 3D printing technology. Face masks are printed in the form of a reusable headpiece to which a separate transparent plastic sheet is attached. It provides the user's eyes and mouth. Although 3D printing provides a way of fighting against the acute shortage of protective kits, it has several disadvantages also. Recently a group of leaders at MIT (Massachusetts Institute of Technology) discussed the limitations of using AM techniques in the making of PPE (personal protective equipment) kits (Badylak et al. 2011). It was noticed that the electrostatic properties of the kits made with conventional methods were hard to be reproduced using 3D printing. The standard tessellation (STL) files required for the print specifications only gives a virtual shape of the model. Small differences in the g-code variables of two 3D printing machines possessing the same STL file will although produce similar appearing models but their functionality will differ. Moreover, the thermoplastic filaments used in additive manufacturing retains moisture which may promote virus growth (Ma 2008). The most challenging factor in the manufacturing of these equipment is the scale of manufacturing. Local 3D printing laboratories produce them at a small scale as it requires several hours to print in on a conventional desktop printer. Although many of these laboratories use multiple printers, the output still remains limited to a dozen masks per printer. Thus, the scale of production of this equipment is presently very low and hence hospitals and 3D printing companies will have to work together in order to solve this manufacturing problem.

11 Applications of 3D Printing Technology

Nowadays numerous medical facilities are now purchasing 3D printers, looking at the enormous potential of 3D printing in the near future. Initially, 3D printing was just seen as a method of gauging the various factors before a surgery was to be executed. But now it is also being widely acknowledged for its potential role in the construction of prosthetic devices, patient education along with clinical training. Biotechnological research mixed with extensive research on 3D printing is being used for the construction of tissues and organs with accurate functionalities. The varied applications of 3D printing in medicine are given in Fig. 6.

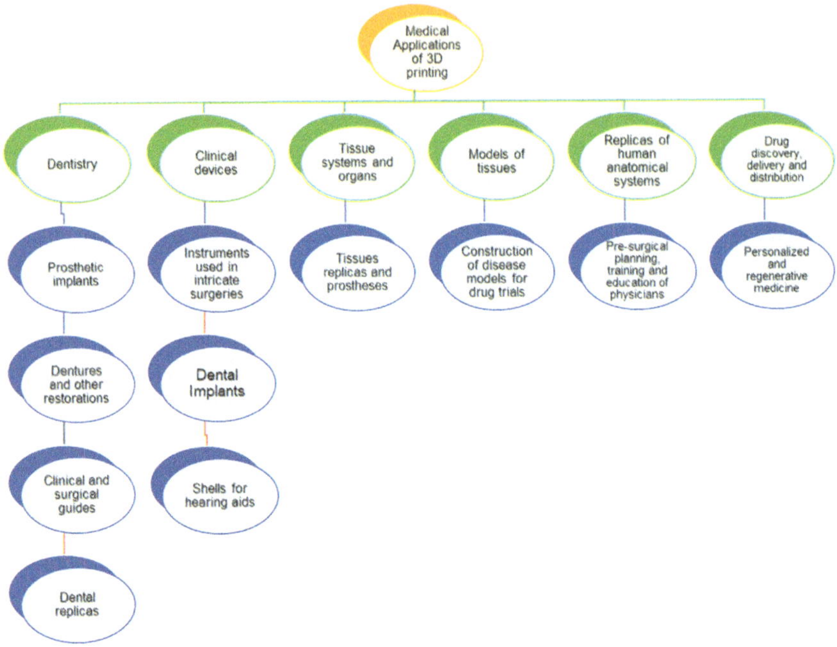

Fig. 6 Applications of 3D printing in medicine

12 Regenerative Medicine

Tissue engineering focuses on mainly two aims, viz., to manufacture tissues and organs for implantation and to manufacture tissue models to study them. Tissue-engineered scaffolds are very important as they provide a structural support for the cells to attach, travel, and multiply (Leong et al. 2003; Ferris et al. 2013; Lee et al. 2016). In additive manufacturing, the cells are either integrated after printing by seeding the seeds into the scaffold or by expanding the cells into the ink formulation before printing. The latter process is also known as 'bioprinting'. Bioprinting can be further classified as scaffold-free and scaffold-based bioprinting (Murphy and Atala 2014).

13 To Eliminate the Prerequisite for Animal Experimentation

Taking into account the digital characteristics and arranging matter in accordance with them, 3D printing is widely used to develop substances with complicated attributes (Ricles et al. 2018; Amir-Aslani and Mangematin 2010; Ma et al. 2018).

The technique of AM can be brought to use in all steps of research in the field of pharmacy inclusive of discovery, development, and delivery of drugs (Fan et al. 2016; Vijayavenkataraman et al. 2018). Drug development also involves extensive research to gauge the extent of its absorption, metabolic activity, excretory activity, toxicity, and the ideal amount of dosage. The 2D monolayer culture complex in which traditionally most of the studies are carried out are not capable of replicating the original 3D microenvironment of the tissues and organs. Therefore, in vitro and in vivo results of the tests display huge variation (Zhang et al. 2016; Kaplowitz 2005; Collins and Varmus 2015; Madathilethu et al. 2018).

14 Personalized Medicines

The trend of personalized medicines in the pharmaceutical industry caught momentum to eradicate the approach of "one size fits all" in the field of treatment (Scoutaris et al. 2018). Traditionally, tablet manufacture occurs in bulk in small distinct doses required for administration in the entirety of the population. The dose requirements of people can vary on an individual basis depending on their genetic makeup, condition of the disease, sex, weight, age, etc. This led to the emergence of personalized medicines and AM played an essential role in augmenting the arena of personalized medicines by making possible the creation of little batches of personalized therapeutics directly on field (Sadia et al. 2018; Jamróz et al. 2017; Vuddanda et al. 2018; Okwuosa et al. 2016; Khaled et al. 2015; Norman et al. 2017). This approach was employed in creating polypill for cardiac diseases containing a formulation of five drugs including hydrochlorothiazide, ramipril, atenolol, pravastatin along with aspirin (Cohen et al. 2009).

3D printing can also be used in making administration fast-paced and improving access to medications. In situations wrought with limitation of time and resources like areas of calamities, accidents, emergency, and military 3D printers could be installed to ensure rapid administration of treatment and therapy (Chung et al. 2014).

15 Imaging of the Head and Neck

The recreation of craniomaxillofacial has always involved the use of 3D printing as a part of extensive planning and preparation before surgery. The facial distortions arising due to resection of tumor are rectified by means of facial reconstruction. These complex operations often involve a huge amount of time which is spent shaping the titanium plates employed in connecting the adjoining bone together, while the patient remains unconscious under the effect of anesthesia (Tam et al. 2012). A lot of time invested in these surgeries can be saved if these plates are shaped before the surgery itself by means of creation of a 3D model of the anatomy of the bone structure. This leads to overall reduction in the exposure of the patient to anesthesia, blood loss is reduced, and the wound is subjected to much less exposure. This approach is also widely employed in orthopedic operations (Imanishi and Choong 2015).

16 Imaging of the Musculoskeletal System

Orthopedics was one of the first areas that employed the use of AM technology. Initially, it was just used in pre-surgical preparations, but subsequently it has been used in designing prosthetic devices. A 6-year-old girl was diagnosed with hereditary multiple exostoses (HME) and was afflicted with a huge osteochondroma in the scapula leading to extreme hindrance in movement and excruciating pain (Schmauss et al. 2014). Owing to the danger of wounding adjacent structures and tissues during the operation, extensive pre-surgical planning was done by making use of a three-dimensional printed replica of the scapula created through CT pictures of the patient in consideration. The replica aided in picturing the tumor aptly and in doing extensive preparations (Schmauss et al. 2014). Employing prosthetic implants made of titanium leads to faster and increased weight-bearing capabilities and increased movement by making use of allograft reconstruction along with autograft correction (Zein et al. 2013).

17 Imaging of Cardiac System

Intraoperative as well as pre-surgical preparation in cardiac surgeries involves vast use of 3D printed cardiac models. This approach largely helps in planning, resolving of issues that might arise amid the surgery and intraoperative strategy making (Schlitter et al. 2012). An augmented understanding of cardiac anatomy is possible through these models which is much better than traditional cardiac imaging.

18 Imaging of the Vascular System

Abdominal aortic aneurysms often involve EVAR (Endovascular aneurysm repair) as one of the major treatment options. This approach is often preferred for its limited invasiveness as compared to open surgery. Although, one drawback of this approach is that sometimes another surgery is required for overhauling of leakages and displacement of grafts. The graft is opted for by means of CT scan pictures but complicated cases still involve unpredictability (Schmauss et al. 2014). The selection of stents in EVAR can be made more accurate by bringing to use AM replicas of abdominal aneurysms by applying information collected from CT scans.

19 Imaging of the Gastrointestinal System

Information collected from the CT scan images of existing donors as well as recipients were used to create AM replicas of the liver, vascular, and biliary system to help with pre-surgical preparations (Waran et al. 2015). A clear, elastic material was used in the construction of the vascular system which augmented better picturing and understanding of the physical arrangement and extent of proximity between the vascular and the biliary arrangement. Employing the analysis made from CT scan of pancreatic carcinoma patients a 3D printed replica inclusive of the pancreas along with the tumor and adjacent structures were constructed (Dong et al. 2020).

20 Imaging of the Genitourinary System

There is extensive research going on in the field of urological surgeries incorporating the use of 3D printers. For kidney stones larger in size than 2 cm an operation is required. This is done through PCNL (Percutaneous nephrolithotripsy). CT scan picturization along with 3D recreation can be used in pre-surgical planning.

21 3D Bioprinting

The recent advancements in the field of bioprinting have rendered the creation of living tissues by using 3D printing. For the purpose of transplantation procedures, different types of tissues including those of the bone, cartilage, and skin have also been prepared. Extensive research is underway to construct real and accurate kidney, liver, and other organs which will be considered as a breakthrough in the field of organ transplantation (Amir-Aslani and Mangematin 2010).

- **3D printed permeable Magnesium scaffolds**-Tu Delft researchers brought to light how a combination of magnesium powder as well as polymer ink which is rheological and suitable can be exuded into scaffolding arrangements found within bone tissues at room temperature and then is subjected to sintering by making use of solvent cast 3D printing technique (Li et al. 2020).
- **3D printed arteries to examine blood pressure-**UW Madison researchers have produced a polarized tube inclusive of ferroelectric materials having a lattice structure that is sinusoidal in nature in addition to silver paste that is bio-compatible on the upper and lower side. It is then enclosed within PDMS for the creation of artificial arteries that are involved in the generation of powerful piezoelectric signals that are then used for monitoring blood pressure (Waimin et al. 2020).
- **3D printed colonoscopy capsules-**Purdue researchers have created a distinct capsule that is capable of sampling bacteria inside the human gut. Fluids can be withdrawn from the entirety of the gastrointestinal system by making use of

hydrogels capable of absorption of bacteria in the gastrointestinal fluids. Samples can be extracted from various parts of the GI tract as the 3D printed capsule travels through it (Abbas et al. 2020).
- **3D printed ciprofloxacin pills enabling modified release along with controlled dosage by FFF technique-**Pakistani researchers created tablets having an intricate combination of ciprofloxacin hydrochloride and PVA filament. These 3D printed tablets have varied fills and motifs aiding in desirable release and controlled dosage (Baumers and Holweg 2019).

22 Education and Training of Physicians

AM is also used to train surgeons. Numerous healthcare institutions are using it to impart training to novice surgeons on complicated surgeries such as those involving congenital cardiac disorders and intricate neurological surgeries. Accurate replicas of anatomy are being considered a breakthrough in the field of neurological studies. Three printing also enables a better hands-on training of neurological and neuroendoscopic operations (Goh et al. 2020).

23 Impact of AI in 3 D Printing

In manufacturing, AM or 3D printing is increasing rapidly because of its ability to produce components with complex features, the industry has attracted a lot of attention from different fields (van Eijnatten et al. 2018). With the help of artificial intelligence and machine learning, 3D-printing technology can be improved and made more efficient. Artificial intelligence and machine learning develop special tools that are able to detect the errors in the model and further rectifying them. It also helps in preventing errors and helps in improving the process of printing. The devices are capable of processing large data, learn and finally implement them in a systematic manner. The algorithm can be learnt more rapidly and further used in 3D printing with the help of artificial intelligence.

The use of artificial intelligence and machine learning in AM helps in the following ways:

- Improving the prefabrication design processes,
- Detects defects and failures,
- Compensation of failures and real-time 3D printing,
- Optimization of cost (workflow),
- Use of machine learning algorithms in photopolymerization and chemical reactions to maximize control, and
- Maintenance is predictive.

Additive manufacturing serves as an umbrella for 3D printing which can be integrated with cloud printing and IoT to manufacture electrical goods and to increase efficiency and decrease costs of the manufacturing industry. 3D printing can be used in cloud manufacturing to gain designing freedom for prototyping. Knowledge of AM can be used in cloud computing when orders placed by customers have to be analyzed, such as to determine the type of 3D printing process to be used, allocating the order to the respective factory, etc.

IoT is a growing field and it is expected that in the coming years it will be manufacturing more and more products and 3D printing technology is going to be used for its modeling and prototyping option. Wireless sensors are a very important component of all IoT applications. In the future where more and more smart devices will be produced, the importance of these sensors will also increase. 3D printing is the best option for the intricate design required for the production of these sensors. Additive manufacturing will be able to recreate the exact same designs with respect to both prototyping and production. In the coming years, 3D printing will be playing a significant role in expanding IoT.

AM anatomical models can bring about more exact treatment arranging, better correspondence, and improved preparation and instruction. The creation of anatomical models includes picture procurement and the remaking of the life structures utilizing CT, picture division, and finally printing of the anatomical models (Radzi et al. 2020; Khan et al. 2018). Bioprinting is a developing field in tissue designing that uses 3D printing cycles to print bio-inks to manufacture tissue-like structures (Mishbak et al. 2019; Menon et al. 2019). ML can help foresee material properties of the different blend creation of the bio-inks just as concocting new scaffold plans that suit specific reason through gaining from a huge information base of materials and plans. Different targets improvement of the printing of bio-ink utilizing ML calculations can be performed. For example (Tan 2018), applied progressive AI to simultaneously streamline material, measure factors, and figure added substance assembling of silicone elastomer through freestyle reversible implanting (Tan 2018). The utilization of ML in 3D printing for building and development can cover different viewpoints including material, plan, and cycle (Lao et al. 2020; Wuest et al. 2016). The quest for new 3D printing materials with specific execution, for example, high pressure and elastic properties, solid split opposition and sturdiness, short setting time, and high setting quality should be done via preparing the ML calculations to identify highlights and examples from a huge information base of accessible material properties.

24 Challenges and Future Prospects of 3D Printing

3D printing has found successful implementation in different fields inclusive of medicine, architectural studies, food and processing, mechanical, chemical, aeronautical, and educational industries. Although, the manufacturing industry is still researching extensively to meet resolve the limitations that 3D printing poses. These can be resolved mostly by incorporating more widespread artificial intelligence techniques. The extra wastage of time, lack of control that is more real time in nature, possible security breach, and bridging the gap between mass production and mass modification and customization can only be met through artificial intelligence. Use of AI can augment the use of 3D printing in situations where there is a need for rapid response and unusual demands. AI can also help in increasing the involvement of customers in large industries by the use of AM. 3D printing can aid in the production of small amounts of requirements but is still not very suitable for mass production (Gunessee et al. 2017).

AI integrated with AM/3D printing can be also utilized in on-site production of materials and repairs within the industry which in turn can stand beneficial for humanitarian supply chains. AI can also serve especially the requirement of exploration of certain motifs, regularities, or irregularities in the manufacturing industry (Waran et al. 2014). Artificial intelligence, blockchain, and 3D printing can be integrated to solve several problems in humanitarian supply chains. AI can work toward improving decision-making while 3D printing improves printing of alternatives on the spot which can go a long way in decongesting supply chains. Their combination can also be used to improve the disaster management system led by the government (Koeppe et al. 2018). Numerical simulations based on physics are considered to be less proficient computationally in contrast to numerical simulations that are data based and make use of machine learning techniques. Stress prognosis of a lattice arrangement through means of a model based on ML takes a mere 0.47 s while a simulation using finite element method takes 5–10 h for completion (Khadilkar et al. 2019). It was also discovered that stress prognosis by means of a convolutional neural network that is data driven took a few milliseconds while the same done using finite element analysis took up to a few minutes (Nagarajan et al. 2018). Training a humongous data set can be computationally prolonged as well as costly. Employing a data-driven artificial neural network model reduced the amount of time and expense devoted to a great extent (Zhang et al. 2016).

3D printing has the potential to bring about revolutionary advancements in the field of health care in the near future. But like every other technology, 3D printing too has its set of challenges that are in need of resolution. The cost of the entire process is one of the major factors in deciding its feasibility in healthcare facilities. Even though the cost of the printers has dipped in recent times, the 3D printers with high resolution optimal for clinical practice are still very expensive (Schmauss et al. 2014). Outsourcing of 3D printers has its different set of drawbacks. Apart from being expensive, it can also lead to hindrance of patient privacy due to off-site translocation of data and shipping might cause an inadvertent delay. The quantity of time and hard work spent to isolate the area of interest from the huge amount of post-process information is another major drawback. The time required for post-processing and analyzing the data in times of emergency can be a major disadvantage.

3D printing is only being applied to unconventional and intricate surgeries involving unique prosthetic implants to date. With the advancement of this technology, it is speculated to be used in traditional normal surgeries as well to contribute to their ease too. To date, there have been no studies extensively comparing the

3D printing technology with the conventional surgical practices. More research is needed to justify and contemplate if the huge expenditure invested in 3D printers will eventually save time and money in the long run. The consistency of the tissues used for these 3D printed models still has some discrepancies as compared to actual human tissues (Koeppe et al. 2018). This is speculated to improve eventually with technological advancements in the field of bioprinting. The development of huge databases involves sharing data which is a requisite for ML algorithms to function. With numerous researchers working extensively toward process development and current materials, data acquisition levels and its pre-processing would encourage data sharing and augment amicable collaboration between various groups in the field of AM. Although, 3D printing is rapidly gaining popularity in the medical and healthcare sector. It offers a highly individual-specific approach and economically feasible, and readily available 3D replicas. The major applications are inclusive of tissue and organ engineering, replicas of intricate anatomy, drug discovery, development and distribution and construction of complex medical devices. Dentistry is one of the major areas where 3D printing is primarily used such as construction of dental replicas, restorations, and clinical guides. Approximately, 85 medical devices created through means of 3D printing are currently commercially available moreover, 99% of hearing aid devices (their shells) are made through 3D printing. Rehearsing on replicas of anatomy before a delicate operation by surgeons and trainees have paved the way for revolutionary surgical techniques.

To date precisely 15 varied kinds of 3D printed tissues have been made available in the market. Although is constantly proving to be a revolutionary technology, it still has a long way to go. There is a need for extensive research for the resolution of the challenges it poses for its full potential to be discovered. More attention needs to be given toward hastening the speed of the 3D printing process along with polishing the resolution, ease of use, and reproducibility. A deeper understanding of proper spatial distribution within complex tissues is of utmost importance to produce exact 3D models. Extensive research is being put into reducing the time and effort invested in post-processing analysis which tends to unnecessarily complicate the process and also increase the cost. There are still some problems regarding the regulatory norms of products created through 3D printing. Extensive research is needed in the field of multi-task-oriented learning that would work toward increasing the dependability of the model that would in turn enable designers to evaluate the performance and capability of AM products before the actual manufacturing. A prognostic model like this would further expedite the processing of digital twins in AM. It opens up new avenues for the expansion of ML and its increased use in the arena of 3D printing.

References

Abbas N, Qamar N, Hussain A, Latif S, Arshad MS, Ijaz QA, Bukhari NI et al (2020) Fabrication of modified-release custom-designed ciprofloxacin tablets via fused deposition modeling 3D printing. J 3D Print Med. https://doi.org/10.2217/3dp-2019-0024

Alomari M, Mohamed F, Basit A, Gaisford S (2015) Personalised dosing: printing a dose of one's own medicine. Int J Pharm 494(2):568–577
Amir-Aslani A, Mangematin V (2010) The future of drug discovery and development: shifting emphasis towards personalized medicine. Technol Forecast Soc Chang 77(2):203–217
Arcaute K, Mann B, Wicker R (2010) Stereolithography of spatially controlled multi-material bioactive poly(ethylene glycol) scaffolds. Acta Biomater 6(3):1047–1054
Badylak S, Taylor D, Uygun K (2011) Whole-organ tissue engineering: decellularization and recellularization of three-dimensional matrix scaffolds. Annu Rev Biomed Eng 13(1):27–53
Bak D (2003) Rapid prototyping or rapid production? 3D printing processes move industry towards the latter. Assembly Autom 23(4):340–345
Bateman N, Leach R (1998) ABC of oxygen: acute oxygen therapy. BMJ 317(7161):798–801
Baumers M, Holweg M (2019) On the economics of additive manufacturing: experimental findings. J Oper Manage 65(8):794–809
Ben, Nacef I, Ben SR, Ouslati I, Khiari K, Ben AN (2014) P163 Diabète et acromégalie. Diabetes Metab 40:A67
Bengio Y, Courville A, Vincent P (2013) Representation learning: a review and new perspectives. IEEE Trans Pattern Anal Mach Intell 35(8):1798–1828
Bruggeman D (2012) Digital apollo: human and machine in spaceflight. MIT Press, Cambridge, MA, xii–362 p. ISBN 9780262134972, $29.95 cloth.; Mindell D (2008) Rev Policy Res 29(5):661–662
Chan V, Collens MB, Jeong JH, Park K, Kong H, Bashir R (2012)
Chen R, Jin Y, Wensman J, Shih A (2016) Additive manufacturing of custom orthoses and prostheses—A review. Addit Manuf 12:77–89
Chung K, Hong D, Kim Y, Yang I, Park Y, Kim H (2014) Preshaping plates for minimally invasive fixation of calcaneal fractures using a real-size 3D-printed model as a preoperative and intraoperative tool. Foot Ankle Int 35(11):1231–1236
Cohen A, Laviv A, Berman P, Nashef R, Abu-Tair J (2009) Mandibular reconstruction using stereolithographic 3-dimensional printing modeling technology. Oral Surg Oral Med Oral Pathol Oral Radiol Endodontol 108(5):661–666
Collins F, Varmus H (2015) A new initiative on precision medicine. N Engl J Med 372(9):793–795
Dong J, Li Y, Lin P, Leeflang MA, van Asperen S, Yu K, Zhou J et al (2020) Solvent-cast 3D printing of magnesium scaffolds. Acta Biomater. https://doi.org/10.1016/j.actbio.2020.08.002
du Boulay B (2001) N.J. Nilsson, artificial intelligence: A new synthesis T. Dean J. Allen and Y. Aloimonos, Artificial Intelligence: Theory and Practice D. Poole A. Mackworth and R. Goebel Computational Intelligence: A Logical Approach S. Russell and P. Norvig, artificial intelligence: a modern approach. Artif Intell 125(1–2):227–232
Erol S, Jäger A, Hold P, Ott K, Sihn W (2016) Tangible industry 4.0: a scenario-based approach to learning for the future of production. Procedia CIRP 54:13–18
Fan Y, Nguyen D, Akay Y, Xu F, Akay M (2016) Engineering a brain cancer chip for high-throughput drug screening. Sci Reports 6(1)
Ferris C, Gilmore K, Wallace G, Panhuis M (2013) Biofabrication: an overview of the approaches used for printing of living cells. Appl Microbiol Biotechnol 97(10):4243–4258
Ghobakhloo M (2018) The future of manufacturing industry: a strategic roadmap toward industry 4.0. J Manuf Technol Manag 29(6):910–936
Goh G, Sing S, Yeong W (2020) A review on machine learning in 3D printing: applications, potential, and challenges. Artif Intell Rev
Gunessee S, Subramanian N, Roscoe S, Ramanathan J (2017) The social preferences of local citizens and spontaneous volunteerism during disaster relief operations. Int J Prod Res 56(21):6793–6808
He W, Xu L (2014) A state-of-the-art survey of cloud manufacturing. Int J Comput Integr Manuf 28(3):239–250
Heaton J (2017) Ian, Goodfellow, Yoshua, Bengio, and Aaron, Courville: deep learning. Genet Program Evolvable Mach 19(1–2):305–307
Imanishi J, Choong P (2015) Three-dimensional printed calcaneal prosthesis following total calcanectomy. Int J Surg Case Reports 10:83–87

Jamróz W, Kurek M, Łyszczarz E, Szafraniec J, Knapik-Kowalczuk J, Syrek K, Paluch M, Jachowicz R (2017) 3D printed orodispersible films with Aripiprazole. Int J Pharm 533(2):413–420

Jurischka C, Dinter F, Efimova A, Weiss R, Schiebel J, Schulz C, Fayziev B, Schierack P, Fischer T, Rödiger S (2020) An explorative study of polymers for 3D printing of bioanalytical test systems. Clin Hemorheol Microcirc 1–28

Kalim, Syed (2020) Maritime economy and gwadar port: a growth catalyst. Policy Perspect 17(1):73

Kang H, Lee S, Ko I, Kengla C, Yoo J, Atala A (2016) A 3D bioprinting system to produce human-scale tissue constructs with structural integrity. Nat Biotechnol 34(3):312–319

Kaplowitz N (2005) Idiosyncratic drug hepatotoxicity. Nat Rev Drug Discovery 4(6):489–499

Khadilkar A, Wang J, Rai R (2019) Deep learning–based stress prediction for bottom-up SLA 3D printing process. Int J Adv Manuf Technol 102(5–8):2555–2569

Khaled S, Burley J, Alexander M, Yang J, Roberts C (2015) 3D printing of five-in-one dose combination polypill with defined immediate and sustained release profiles. J Control Release 217:308–314

Khan Z, Kahin K, Rauf S, Ramirez-Calderon G, Papagiannis N, Abdulmajid M, Hauser C (2018) Optimization of a 3D bioprinting process using ultrashort peptide bioinks. Int J Bioprint 5(1)

Koeppe A, Hernandez Padilla C, Voshage M, Schleifenbaum J, Markert B (2018) Efficient numerical modeling of 3D-printed lattice-cell structures using neural networks. Manuf Lett 15:147–150

Lakafosis V, Rida A, Vyas R, Yang L, Nikolaou S, Tentzeris M (2010) Progress towards the first wireless sensor networks consisting of inkjet-printed, paper-based RFID-enabled sensor tags. Proc IEEE 98(9):1601–1609

Lao W, Li M, Wong T, Tan M, Tjahjowidodo T (2020) Improving surface finish quality in extrusion-based 3D concrete printing using machine learning-based extrudate geometry control. Virtual Phys Prototyp 15(2):178–193

LeCun Y, Bengio Y, Hinton G (2015) Deep learning. Nature 521(7553):436–444

Lee S, Lee D, Yoon T, Kim H, Jo H, Park J, Lee J, Kim W, Kwon I, Park S (2016) Surface modification of 3D-printed porous scaffolds via mussel-inspired polydopamine and effective immobilization of rhBMP-2 to promote osteogenic differentiation for bone tissue engineering. Acta Biomater 40:182–191

Lee K, Park K, Hwang J, Lee M, Yoon D, Ryu H, Jung H, Park K, Kim J, Park S, Kim S, Chun Y, Choi W, Lee J (2018) Inhibition of STAT5A promotes osteogenesis by DLX5 regulation. Cell Death Dis 9(11)

Leong K, Cheah C, Chua C (2003) Solid freeform fabrication of three-dimensional scaffolds for engineering replacement tissues and organs. Biomaterials 24(13):2363–2378

Li J, Long Y, Yang F, Wei H, Zhang Z, Wang Y, Wang X et al (2020) Multifunctional artificial artery from direct 3D printing with built-in ferroelectricity and tissue-matching modulus for real-time sensing and occlusion monitoring. Adv Func Mater 2002868

Livingston E, Desai A, Berkwits M (2020) Sourcing personal protective equipment during the COVID-19 pandemic. JAMA 323(19):1912

Löffler C, Westkämper E, Unger K (2011) Method for analysis and dynamism of factory structure in automotive manufacturing. Robot Comput Integr Manuf 27(4):741–745

Low Z, Chua Y, Ray B, Mattia D, Metcalfe I, Patterson D (2017) Perspective on 3D printing of separation membranes and comparison to related unconventional fabrication techniques. J Membr Sci 523:596–613

Ma P (2008) Biomimetic materials for tissue engineering. Adv Drug Deliv Rev 60(2):184–198

Ma X, Liu J, Zhu W, Tang M, Lawrence N, Yu C, Gou M, Chen S (2018) 3D bioprinting of functional tissue models for personalized drug screening and in vitro disease modeling. Adv Drug Deliv Rev 132:235–251

Madathilethu J, Roberts M, Peak M, Blair J, Prescott R, Ford J (2018) Content uniformity of quartered hydrocortisone tablets in comparison with mini-tablets for paediatric dosing. BMJ Paediatr Open 2(1):e000198

Mau R, Nazir J, John S, Seitz H (2019) Preliminary study on 3D printing of PEGDA hydrogels for frontal sinus implants using digital light processing (DLP). Curr Direct Biomed Eng 5(1):249–252

Menon A, Póczos B, Feinberg A, Washburn N (2019) Optimization of silicone 3d printing with hierarchical machine learning. 3D Print Addit Manuf 6(4):181–189
Mishbak HH, Cooper G, Bartolo P (2019) Development and characterisation of a photocurable alginate bioink for 3D bioprinting. Int J Bioprint 5(2):12
Murphy S, Atala A (2014) 3D bioprinting of tissues and organs. Nat Biotechnol 32(8):773–785
Nagarajan H, Mokhtarian H, Jafarian H, Dimassi S, Bakrani-Balani S, Hamedi A, Coatanéa E, Gary Wang G, Haapala K (2018) Knowledge-based design of artificial neural network topology for additive manufacturing process modeling: a new approach and case study for fused deposition modeling. J Mech Des 141(2)
Norman J, Madurawe R, Moore C, Khan M, Khairuzzaman A (2017) A new chapter in pharmaceutical manufacturing: 3D-printed drug products. Adv Drug Deliv Rev 108:39–50
Norvig P (2003) PowerPoint: shot with its own bullets. Lancet 362(9381):343–344
Okwuosa T, Pereira B, Arafat B, Cieszynska M, Isreb A, Alhnan M (2016) Fabricating a shell-core delayed release tablet using dual FDM 3D printing for patient-centred therapy. Pharm Res 34(2):427–437
Pandey S, Gupta B, Nahata A (2013) Terahertz plasmonic waveguides created via 3D printing. Opt Express 21(21):24422
Peyre P, Dal M, Pouzet S, Castelnau O (2017) Simplified numerical model for the laser metal deposition additive manufacturing process. J Laser Appl 29(2):022304
Popov A, Malferrari S, Kalaskar D (2017) 3D bioprinting for musculoskeletal applications. J 3D Print Med 1(3):191–211
Radzi S, Tan H, Tan G, Yeong W, Ferenczi M, Low-Beer N, Mogali S (2020) Development of a three-dimensional printed heart from computed tomography images of a plastinated specimen for learning anatomy. Anat Cell Biol 53(1):48
Reddy R, Ram N (2020) Data aggregation and precedence by delay sensitivity (DAP-DS): Data transmission over wireless body sensor networks. Microprocess Microsyst 77:103165
Ricles L, Coburn J, Di Prima M, Oh S (2018) Regulating 3D-printed medical products. Sci Transl Med 10(461):eaan6521
Rochman D (2020) Make or buy: case study of 3D printing spare parts adoption. Int J Psychosoc Rehabil 24(02):2940–2945
Sadia M, Arafat B, Ahmed W, Forbes R, Alhnan M (2018) Channelled tablets: An innovative approach to accelerating drug release from 3D printed tablets. J Control Release 269:355–363
Schlitter A, Michalski C, Kleeff J, Dobritz M, Esposito I (2012) Unklare zystische Raumforderung im Pankreasschwanz. Gastroenterologie up2date 08(04):253–257
Schmauss D, Haeberle S, Hagl C, Sodian R (2014) Three-dimensional printing in cardiac surgery and interventional cardiology: a single-centre experience. Eur J Cardiothorac Surg 47(6):1044–1052
Scoutaris N, Ross S, Douroumis D (2018) 3D printed "Starmix" drug loaded dosage forms for paediatric applications. Pharm Res 35(2)
Shen X, Naguib H (2019) A robust ink deposition system for binder jetting and material jetting. Addit Manuf 29:100820
Smoczok M, Starszak K, Starszak W (2020) 3D printing as a significant achievement for application in posttraumatic surgeries—a literature review. Curr Med Imaging Former Curr Med Imaging Rev 16
Stansbury J, Idacavage M (2016) 3D printing with polymers: challenges among expanding options and opportunities. Dent Mater 32(1):54–64
Tam M, Laycock S, Bell D, Chojnowski A (2012) 3-D printout of a DICOM file to aid surgical planning in a 6-year-old patient with a large scapular osteochondroma complicating congenital diaphyseal aclasia. J Radiol Case Reports 6(1)
Tan K (2018) The framework of combining artificial intelligence and construction 3D printing in Civil Engineering. MATEC Web of Conf 206:01008
Tao F, Zhang L, Liu Y, Cheng Y, Wang L, Xu X (2015) Manufacturing service management in cloud manufacturing: overview and future research directions. J Manuf Sci Eng 137(4)
Tibbits S (2016) Additive hybrids. 3D Print Addit Manuf 3(1):1

Tiwari S, Pande S, Agrawal S, Bobade S (2015) Selection of selective laser sintering materials for different applications. Rapid Prototyp J 21(6):630–648

Tofail S, Koumoulos E, Bandyopadhyay A, Bose S, O'Donoghue L, Charitidis C (2018) Additive manufacturing: scientific and technological challenges, market uptake and opportunities. Mater Today 21(1):22–37

van Eijnatten M, van Dijk R, Dobbe J, Streekstra G, Koivisto J, Wolff J (2018) CT image segmentation methods for bone used in medical additive manufacturing. Med Eng Phys 51:6–16

Vijayavenkataraman S, Yan W, Lu W, Wang C, Fuh J (2018) 3D bioprinting of tissues and organs for regenerative medicine. Adv Drug Deliv Rev 132:296–332

Vuddanda P, Alomari M, Dodoo C, Trenfield S, Velaga S, Basit A, Gaisford S (2018) Personalisation of warfarin therapy using thermal ink-jet printing. Eur J Pharm Sci 117:80–87

Waimin JF, Nejati S, Jiang H, Qiu J, Wang J, Verma MS, Rahimi R (2020) Smart capsule for non-invasive sampling and studying of the gastrointestinal microbiome. RSC Adv 10(28):16313–16322. https://doi.org/10.1039/c9ra10986b

Wang Y, Blache R, Xu X (2017) Selection of additive manufacturing processes. Rapid Prototyp J 23(2):434–447

Waran V, Narayanan V, Karuppiah R, Pancharatnam D, Chandran H, Raman R et al (2014) Injecting realism in surgical training—Initial simulation experience with custom 3D models. J Surg Educ. 71(2):193–197

Waran V, Narayanan V, Karuppiah R, Thambynayagam H, Muthusamy K, Rahman Z, Kirollos R (2015) Neurosurgical endoscopic training via a realistic 3-dimensional model with pathology. Simul Healthc: J Soc Simul Healthc 10(1):43–48

Wu D, Greer M, Rosen D, Schaefer D (2013) Cloud manufacturing: strategic vision and state-of-the-art. J Manuf Syst 32(4):564–579

Wuest T, Weimer D, Irgens C, Thoben K (2016) Machine learning in manufacturing: advantages, challenges, and applications. Prod Manuf Res 4(1):23–45

Xia F, Yang L, Wang L, Vinel A (2012) Internet of things. Int J Commun Syst 25(9):1101–1102

Xu X (2012) From cloud computing to cloud manufacturing. Robot Comput Integr Manuf 28(1):75–86

Yap Y, Tan Y, Tan H, Peh Z, Low X, Yeong W, Tan C, Laude A (2017) 3D printed bio-models for medical applications. Rapid Prototyp J 23(2):227–235

Zein N, Hanouneh I, Bishop P, Samaan M, Eghtesad B, Quintini C, Miller C, Yerian L, Klatte R (2013) Three-dimensional print of a liver for preoperative planning in living donor liver transplantation. Liver Transpl 19(12):1304–1310

Zhang Y, Yue K, Aleman J, Mollazadeh-Moghaddam K, Bakht S, Yang J, Jia W, Dell'Erba V, Assawes P, Shin S, Dokmeci M, Oklu R, Khademhosseini A (2016) 3D bioprinting for tissue and organ fabrication. Ann Biomed Eng 45(1):148–163

3D Printing Technology for Fighting COVID-19 Pandemic

Rohin Shyam, Pearlin Hameed, P. Suya Prem Anand, Loganathan Rangasamy, Arunkumar Palaniappan, and Geetha Manivasagam

Abstract This chapter deals with the emerging applications of 3D printing (3DP) technologies to tackle the recent pandemic, the Corona viral disease (COVID-19). The chapter throws light on the role of 3DP technologies and other allied hybrid technologies for the development of novel products to satiate the shortage of personal protective equipment (PPE) such as face shields, masks, eye protection devices, ventilator tubes, and other medical devices needed to tackle COVID-19. It also explicates the hybrid additive processes required to fabricate novel metal and ceramic based biomedical implants with inbuilt antimicrobial and antiviral properties. Also, in vitro lung tissue models, especially based on 3D bioprinting technology for the screening of novel anti-SARS-CoV-2 (Severe Acute Respiratory Syndrome Coronavirus 2) therapies are also elaborated in-depth. Finally, 3DP technologies based bespoke drug delivery devices for personalized and on-demand drug dosing, complex drug release profiles, and polypills are discussed. To conclude, this chapter emphasizes the role of 3DP technologies in the development of novel emerging applications like antiviral property enriched biomedical implants, fabrication of PPE, in vitro lung tissue models, and finally personalized drug delivery devices, which could go a long way in tackling COVID-19 in an efficient manner.

Keywords COVID-19 · 3D printing · Personal protective equipment · Social care

1 Introduction

Corona virus disease (COVID-19) caused by the novel severe acute respiratory syndrome coronavirus 2 (SARS-CoV-2) which is declared as a pandemic by WHO (World Health Organization) is spreading rapidly across the globe. As of October

R. Shyam · P. Hameed · P. Suya Prem Anand · L. Rangasamy · A. Palaniappan (✉) · G. Manivasagam (✉)
Centre for Biomaterials, Cellular and Molecular Theranostics (CBCMT), Vellore Institute of Technology, Vellore, India
e-mail: arunkumar.p@vit.ac.in

G. Manivasagam
e-mail: geethamanivasagam@vit.ac.in

K. Sandhu et al. (eds.), *Emerging Applications of 3D Printing During CoVID 19 Pandemic*, Lecture Notes in Bioengineering, https://doi.org/10.1007/978-981-33-6703-6_5

9th, 2020, total number of people infected with SARS-Cov-2 has reached 36,754,395 and the number of deaths have reached 1,064,838 around the globe (WHO Coronavirus Disease (COVID-19) Dashboard | WHO Coronavirus Disease (COVID-19) Dashboard n.d.). Researchers across the globe are working towards developing novel solutions for combating this pandemic. This includes developing efficient personal protective equipment (PPEs), better therapeutic agents, and novel custom-made drug delivery systems to provide symptomatic relief, as well as to reduce the viral load, better tissue models as drugs/vaccines screening platform to determine the therapeutic efficacy and to understand the mechanism of interaction of viral particles with various tissues.

1.1 Importance of 3DP Technologies in Biomedical Applications

3DP is an additive manufacturing (AM) process which involves layer-by-layer deposition of materials such as plastics, metals, and ceramics to generate intricate objects in 3D space. 3DP is a rapidly growing field of research which has a huge impact in a wide area of applications including automobiles, industrial machines, consumer goods, architecture, consumables, defense sector, and more recently in the medical and dental fields. 3DP technologies can have a huge impact in combating this pandemic through the development of efficient PPEs, better medical devices with antiviral and antimicrobial properties, improved tissue-mimicking drug screening platforms, and finally innovative drug delivery systems to deliver multiple drugs simultaneously with appropriate drug release kinetics.

Out of the plethora of additive manufactured products, a substantial section amounting to 15–16%, is earmarked by the healthcare and medical industry as shown in Fig. 1 (Jiménez et al. 2019; Verhoef et al. 2018). Before the emergence of additive manufacturing in industrial sectors, all PPEs and biomedical implants were marketed

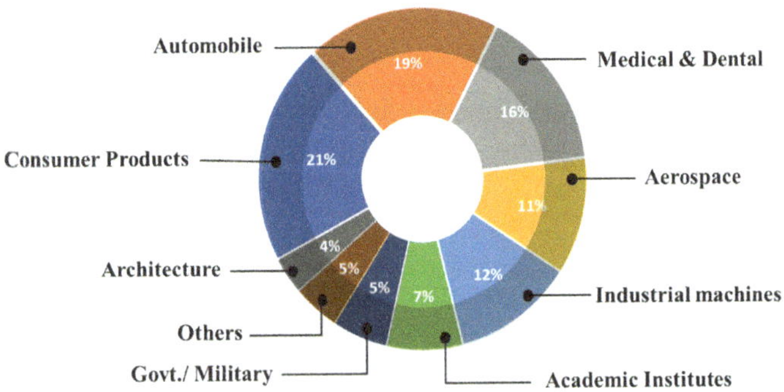

Fig. 1 Distribution of additively manufactured products in different sectors (Jiménez et al. 2019; Verhoef et al. 2018)

solely off-the-shelf in standard and limited size range. 3DP broke new ground by offering in-house production of on-demand, custom-made implants with an opportunity to fabricate complex designs. The incorporation of metal-based 3DP technology such as Selective Laser Melting (SLM), Selective Laser Sintering (SLS), Electron beam melting (EBM), Direct Metal Laser Sintering (DMLS) by manufacturing industries has enabled the rapid fabrication of tailor-made products. In the initial months of the COVID-19 pandemic, global and local supply chains were disrupted resulting in the inability of medical equipment manufacturing industries to meet the demands in supplying critical medical care products. 3DP came to the aid (Shyam and Ramalingam n.d.), especially in the rapid manufacturing of ventilator valves. 3DP played a major role in closing the gap between supply and demand in a time of crisis.

2 3DP for the Fabrication of PPEs

The basic and most important PPEs for combating COVID-19 include face mask, ventilator valve, face shield, and other medical instruments. In order to achieve the ever-expanding need for PPEs, 3DP technologies are being implemented to build a customized design mask (N95) and face shield products (Ishack and Lipner 2020; Swennen et al. 2020; Amin et al. 2020). The World Health Organization (WHO) attributed a spike in the number of COVID-19 positive cases to a shortage of medical equipment and hospital beds. While beds remain a critical challenge for hospitals, the more dire need is to manufacture critical medical equipment, specifically testing kits, masks, and face shields in large numbers, not only for patients, but also for healthcare professionals dealing with COVID-19. With the assistance of 3DP technologies, the rapid manufacturing of critical medical equipment is feasible. This will increase the focus of 3DP technology in the medical field internationally (Tino et al. 2020; Wesemann et al. 2020). In the following sections, the various PPEs such as face mask, face shield, critical medical equipment like ventilator valve, nasal swabs, and innovative products such as non-contact hook manufactured using 3DP technologies are discussed.

2.1 *Face Mask and Face Shield*

Currently, N95 mask is a highly protective mask used for mitigating the spread of COVID-19 due to its enhanced filtering capability of 0.3 μm airborne particles and its sealing advantage which inhibit small dust particles around the edges of the mask, when compared to the cloth and paper mask commercially available in the market which are sold in specific sizes and are uncomfortable to wear for extended durations by healthcare workers and others. 3DP technology is employed to improve the comfort and fitness as shown in Fig. 2a, b. In 3DP, there is a scope for the customization of the design based on the facial features such as nose length, jawline, and chin

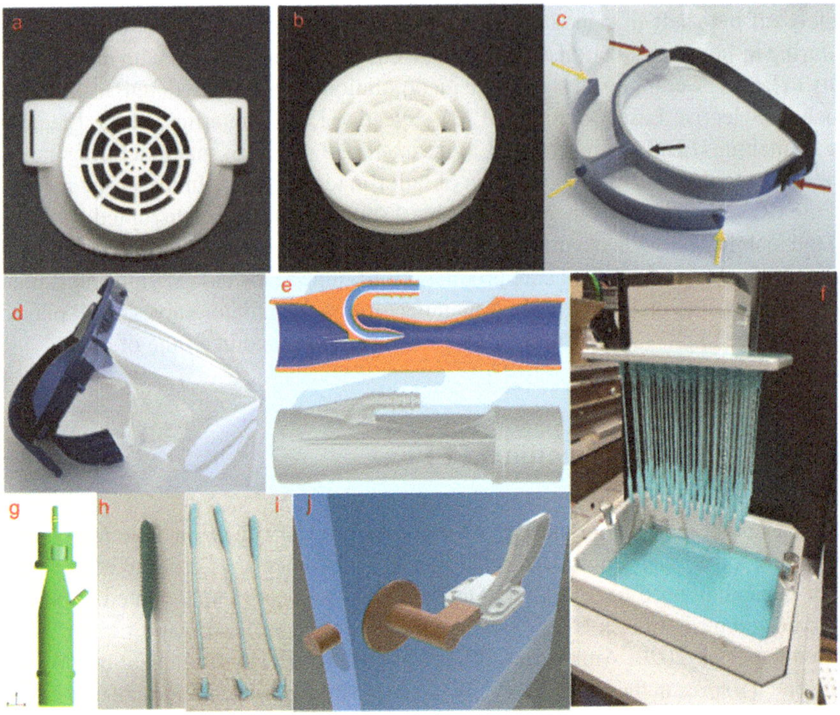

Fig. 2 **a** 3D printed face mask, **b** filter membrane support (Swennen et al. 2020), **c** Face shield frame, **d** face shield (Amin et al. 2020), **e** cross section of ventilator valve (Tino et al. 2020), **f** 3D printed nasopharyngeal swab for sampling COVID-19 and other respiratory viruses (Gallup et al. 2020), **g** CAD model of ventilator valve, **h, i** side view of head of tip swab with tip and nobs (Gallup et al. 2020), **j** non-contact hook (Tino et al. 2020)

arc. Recently, Fused Deposition Modeling (FDM) process was used to manufacture an N95 design mask with Acrylonitrile Butadiene Styrene (ABS) as a processing material. This improves the contact pressure for sealing compared to other respiratory masks. The N95 mask also contains a filtration section which is made up of polypropylene, a material that is lightweight, rigid, and has cyclic load resistant property. However, it is difficult to print using 3DP technology because of distortion. To solve this issue, a thermoplastic Styrene-ethylene-butylene-styrene (SEBS) elastomer like material is blended with polypropylene to improve the printability and flexible nature of the material (Swennen et al. 2020; Tino et al. 2020). Aside from face masks, face shields are another PPE that are intended to keep small droplets from a sneeze or cough landing on the orifices on the face. Face shield can be easily printed with the support of 3DP machines for the healthcare workers in the hospitals. A face shield is made up of polymers like polycarbonate, Polyvinyl Chloride (PVC), and polyester, which have the properties of being lightweight, and transparent material (Ishack and Lipner 2020; Wesemann et al. 2020) The face shield consists of two sections, a reusable headpiece frame manufactured by the 3DP technique and a transparent plastic sheet attached to the headpiece frame as shown in Fig. 2c, d.

2.2 *Ventilator Valve and Nasopharyngeal Swab*

A ventilator valve is used to supply oxygen at a certain concentration in the respiratory system. During the outbreak of COVID-19, patients reported difficulty in breathing and as a result were placed on ventilators. Globally, as the number of COVID-19 cases increased, there rose a huge demand for ventilators for critically ill-patients. Due to global lockdown, medical equipment manufacturers were unable to supply critical parts such as ventilator valves in time. While ventilator parts are manufactured under strict regulations, life-threatening circumstances warranted the need to supersede protocol and bolster the local supply of such critical parts, which were primarily manufactured via 3DP. The key advantage at the time of such a crisis was that the design of an acceptable prototype was shared as an open access, enabling the printing of such parts across the world by hospitals and manufacturers with access to 3D printers. A cross-sectional view of the ventilator valve is shown in Fig. 2e, g. This valve has been made up of different biomaterials such as silicone rubber, polycarbonate, polyamide, and stainless steel. The ventilator valves are produced with the filament extrusion process like fused deposition modeling (FDM) and polymer bed fusion process (Tino et al. 2020). Aside from the manufacturing of valves and other spare parts for medical devices, a major challenge was access to nasopharyngeal swabs for COVID-19 testing. The halting of global supply chain restricted the availability of testing swabs. Herein, stereolithography, a form of 3DP (Fig. 3b), was employed in the mass manufacture of swabs as can be seen in Fig. 2f, i. The ability to mass produce swabs at point of contact increases testing capabilities at facilities which were directly affected by the inability of medical equipment manufacturers to supply such critical devices.

2.3 *Non-contact Hook*

The transmission of the virus can occur through contact with surfaces such as hospital doors and handles which can be subjected to multiple physical contacts by patients infected with the disease. Engineers and innovators designed a door handle that would avoid direct contact with hospital doors and other surfaces in high contact areas. The door handle is also printed with the support of a 3DP machine, where the transmission of virus spread is limited by using the non-contact hook in the doors. Figure 2j shows the 3D printed non-contact hook fixed to the doors. As the coronavirus transfer through the contact surfaces, care should be taken to avoid direct contact with the affected zone. Especially this type of handle can be used in public areas such as hospitals, malls, airports, and railway stations where control the

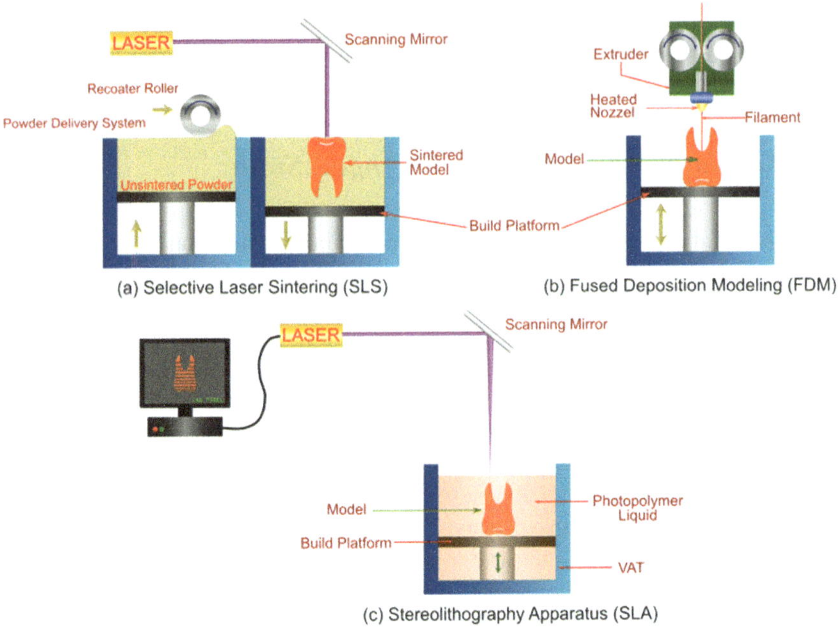

Fig. 3 Example of Additive manufacturing techniques used in the generation of PPE's during COVID-19 lockdown

spread of the virus from the infected person is challenging (Tino et al. 2020). It is evident that 3DP played a major role during the pandemic. One of the challenges that remain to be addressed is the regulations pertaining to 3D printed parts for medical devices. Currently, there are strict guidelines and regulations related to the safety of the manufactured devices. Additionally, the supply and demand of critical medical care devices also play a crucial role. As 3DP technology evolves, these regulations and the designs for parts that can be manufactured via 3DP could be available over public domain without compromising safety.

3 Hybrid Additive Manufacturing for PPEs

The challenges associated with using conventional manufacturing processes for manufacturing PPEs and other products in medical fields include the complex nature of the product both in terms of geometry and design, and complex production cycle, resulting in an increased manufacturing cost. By introducing the AM process in this field will help in overcoming the above-said challenges, by enabling a reduction in the cost involved, lightweight product, multi-material product, reduces manual labor cost, lower tool cost, and combination of the different manufacturing processes in a single machine (Jiménez et al. 2019; Merklein et al. 2016; Karunakaran et al.

2010). Additionally, additive manufacturing, also termed rapid prototyping, has a key advantage in the ability to manufacture prototypes and objects with high precision and agility. This characteristic enables the ease in deploying this technology in an emergency, such as those witnessed during the COVID-19 pandemic.

3.1 Hybrid Additive Manufacturing Process for PPEs

Generally, the Fused Deposition Modeling (FDM) and Stereolithography process are used to fabricate the face mask, face shield, and ventilator valve, etc. Polylactic Acid (PLA) material remains the preferred choice to Acrylonitrile Butadiene Styrene (ABS) material in the FDM process due to biodegradability of PLA (Rodríguez-Panes et al. 2018). In the FDM process, the thermoplastic polymer material in solid or coil filament is heated through the nozzle as shown in Fig. 3b. The nozzle is placed with the extrusion head, where it can be moved along the x and y-axis direction controlled by a DC motor and limits switches. The extruded filament is deposited on the platform that can be moved in the vertical direction. The process is repeated to build the product completely layer-by-layer. The FDM process is preferred to the stereolithography due to the low cost, limited post-processing, and easy printing capability (Ćwikła et al. 2017). Stereolithography consists of a vat arrangement, where thermosetting plastic materials like resins are filled in a vat (container). A photon source is used to polymerize the resin material only in the desired location based on the inputs from the 3D model fed into the printer system. As a result of polymerization, the liquid phase resin is converted into a 3D solid product of desired geometry as shown in Fig. 3c. The process of integrating additive, subtractive, and inspection processes in a single machine is developed to minimize the cost of material, production time, and improve the surface quality. The operations involved in the machining process are controlled by a novel algorithm. Now, the Fused Filament Fabrication and Computer Numerical Control (CNC) milling machines are coupled to produce complex part geometries (Zhu et al. 2014; Choudhari and Patil 2016).

4 3D Printing for Biomedical Implants

The biomedical application of additive manufactured products comprises of the following but are not limited to prosthetic implants, craniofacial plate implants, dental implants, spinal and sternums implants, hip and knee implants, stents, and surgical guides as shown in Fig. 4.

Among the abovementioned implants, clinical trials and successful implantation of the following products have already been achieved. Craniofacial plates/mesh, made of Ti-6Al-4V, fabricated using EBM technology has shown promising results for skull reconstruction (Wang and Qi 2020; Park et al. 2016), Ti-6Al-4V jaw, dental and subperiosteal implant fabricated using DMLS technology demonstrated successful

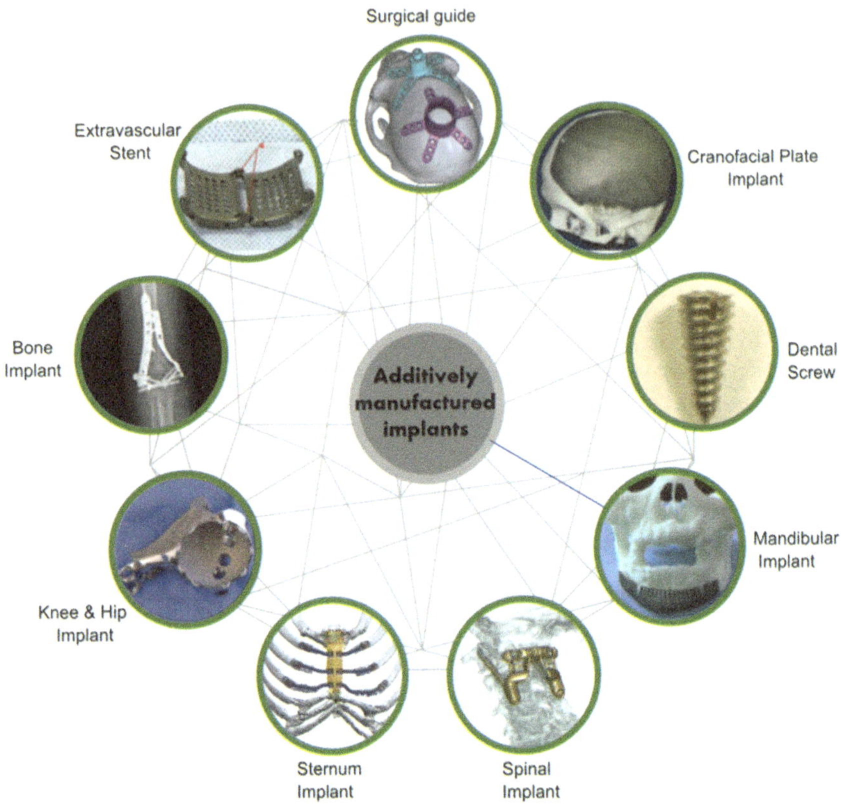

Fig. 4 Additively manufactured implants (clockwise from top). Surgical guide (Chen et al. 2017), craniofacial plate implant (Jardini et al. 2016), dental screw (Tedesco et al. 2017) and mandibular implant (Chen et al. 2017), spinal implant (Phan et al. 2016) and sternum implant (Dzian et al. 2018), hip (Wong et al. 2015) and knee implant, bone implant (Shuang et al. 2016), extravascular stent (Wang et al. 2019). Reprinted under copyright CCBY-NC-ND 4.0 license

osteointegration circumventing bone regeneration therapies (Malyala et al. 2017; Mangano et al. 2020). Titanium spinal implant fabricated using EBM and DMLS have enabled reconstruction of cervical spine (Xu et al. 2016; Willemsen et al. 2019), titanium sternal implant fabricated using DMLS offers progressive treatment for sternal cancer or total sternectomy (Dzian et al. 2018), Titanium hip implant fabricated using EBM for successful Hemipelvic Arthroplasty (Zhao et al. 2018).

4.1 *Surface Functionalization on AM Parts with Antibacterial Effect*

Majority of metal AM applications are devoted to orthopedics due to the high compression and tensile nature of the hard tissue. For soft tissue regeneration, polymer extrusion method and 3D bioprinting are looked upon. To increase the multi-functionality of implant, functionalizing with antibodies allow them to mimic the natural immune mechanism of the body to fight against infection. Achieving antibacterial surfaces or infection resistant surface on implants are under serious consideration as they drastically reduce the post-operative infections, biomedical device-associated infections (BAIs) and reduce the antibiotic dosage. Researchers have experimented with various processes to modify additively manufactured surfaces. Simple acid/alkali etching, or anodization followed by heat treatment have been shown to improve different aspects of additively manufactured samples. For example, Ti-6Al-4V porous samples fabricated using SLM upon acid/alkali treatment enhance the apatite formation indicating a suitable platform for bone mineral deposition. While anodization followed by heat treatment improves the cell proliferation, attachment, and better mechanical stability than acid/alkali treated samples (Yavari et al. 2014a). However, in terms of mechanical properties, the untreated sample outplayed the acid/alkali with or without heat treatment. This was due to loss of mass during acid/alkali treatment (Yavari et al. 2014b). In another study, electrochemical anodization of additively manufacture Ti-6Al-4V resulted in the formation of nanotubes on the surface. These nanotubes can then be used as a drug carrier for delivering other essential drugs based on the need of the patient post-implantation. In addition to being drug carriers, these nanotubes have shown to upregulate osteogenic genes such as osteocalcin & collagen 1 and enhanced ALP production in comparison to as built or unmodified Ti-6Al-4V substrate. Li et al. were successful in loading an MTAN inhibitor (which targets bacteria metabolism responsible for the formation of biofilm) inside nanotubes grown on the surface of the additively manufactured substrate (Li et al. 2020). Apart from nanotube, microcraters and wire structures are also experimented to be used for encapsulation of drugs. Hydrothermal treatment on titanium substrate develops a micro/nano crater structure of titanate layer. Jia et al. were able to encapsulate silver nanoparticles by physically crosslinking silk with polydopamine after creating a micro/nanocrator structure using hydrothermal treatment on EBM printed porous Ti-6Al-4V substrate (Jia et al. 2018). Similarly, using cost-effective and facile methods such as Electrophoretic deposition (EPD), Surmeneva et al. improved antibacterial property on Ti-6Al-4V substrate (without etching) produced by EBM. The deposition of silver nanoparticles on the surface of fabricated substrate empowered antibacterial effect. Interestingly, upon combining calcium phosphate nanoparticles with silver nanoparticles and coating on Ti-6Al-4V substrate, further enhanced antibacterial effect (Surmeneva et al. 2019). Apart from antibacterial property, calcium phosphate coating is also carried out to promote the osteoconductive property of orthopedic implants. Calcium phosphate nanoparticles deposited on additive manufacturing Ti-6Al-4V substrate using EPD show increased ALP activity (Chudinova et al. 2019).

Zwitterionic surfaces have been shown to resist bacterial adhesion. Surface zwitterionization on additively manufactured parts can be a promising modification technique in future to reduce implant infection. EBM manufactured Ti-6Al-4V parts after functionalization with zwitterion ion show a significant reduction in bacterial adhesion of more than 97% with complete prevention of biofilm formation (Rodriguez-Palomo et al. 2016). A commendable example of surface functionalized additive manufactured product is demonstrated by work carried out by Guo et al., where porous Ti-6Al-4V scaffolds printed using SLM are coated with TiCu/Ti–Cu–N via arc ion plating technique. The TiCu/Ti–Cu–N coating on the implant allowed the formation of antibacterial and bioactive coating. The in vitro experiments revealed, surface functionalized promote stem cell proliferation, adhesion and upregulated p38 (a protein responsive for stress stimulation) expression. While in vivo results showed a clear effect of TiCu/Ti–Cu–N coating in promoting bone regeneration (Guo et al. 2020). Guan et al. carried a study on a similar theory, they fabricated Ti-6Al-4V substrate using DMLS, modified the surface, and coated it with antibacterial drug (Guan et al. 2016). The etching of metal surface using HF allows removal of unmelted or weakly bound particles and increases surface roughness to consent better adhesion of antibacterial coating. The deposition of lysosome and subsequent layering with hyaluronic acid (HA) and chitosan (CS) loaded with minocycline hydrochloride (MH) on the etched surface allowed the implant to inhibit biofilm growth as shown in Fig. 5.

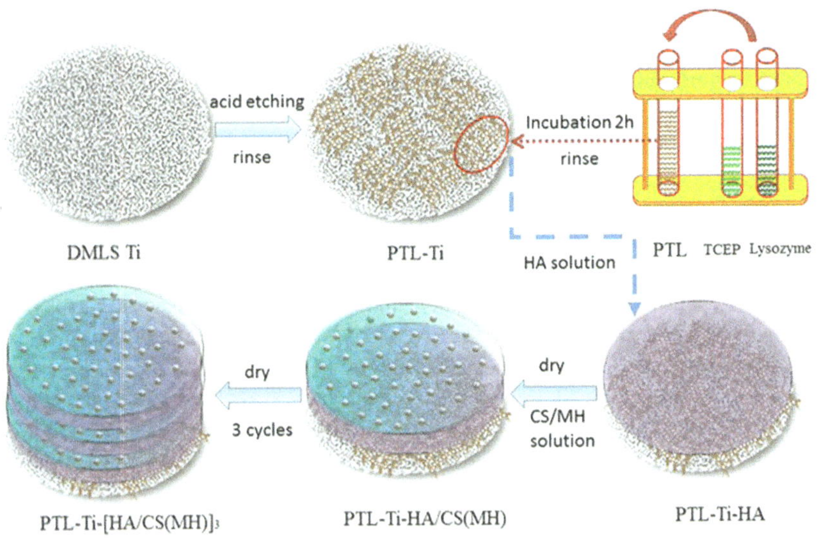

Fig. 5 Surface functionalization of Ti-6Al-4V substrate produced by DMLS to produce an antibacterial multilayer coating by sustained release of minocycline hydrochloride (MH). Adapted from (Guan et al. 2016) under copyright CC BY 4.0 licence

In a similar approach using layer-by-layer depositing on SLM printed porous gyroid Ti-6Al-4V sample, Yavari et al. functionalized the samples using gelatin and chitosan to load BMP-2 or vancomycin respectively. This simple dip process results in a multifunctional implant with antibiotic and osteoinductive properties corresponding to vancomycin and BMP-2, respectively. A sustained release of both the drugs was observed for 2–3 weeks with a strong antibacterial profile. Additionally, the in vitro investigation of multifunctional implants revealed a two-fold and four-fold increase in ALP and mineralization of stem cells respectively (Yavari et al. 2020). Plasma Electrolytic Oxidation (PEO) or Microarc Oxidation (MAO) or Electrolytic Plasma Oxidation (EPO), is an electrochemical surface treatment technique which generates a plasma discharge to coat the substrate with an oxide layer when immersed in an electrolyte. An upside to using PEO is that it modifies the entire surface/geometry without limiting to the line of sight surfaces and modifies the dense surface into interconnected micro or nanoporous structure. Studies on functionalizing of EBM and SLM printed Ti-6Al-4V using PEO have been carried out to produce Ag nanoparticles, TiO_2, and CaP/TiO_2 oxide layer. The coating of silver nanoparticles on SLM printed Ti-6Al-4V has been shown to release silver ions for a sustainable period of 28 days providing an antibacterial environment to inhibit biofilm formation against S. aureus (van Hengel et al. 2017). While TiO_2 surface modification promotes bone formation inside and around the scaffold and allows better bone implant anchorage due to the modified topology. Scientists have augmented the surface functionalization process by attaching antibacterial and anticoagulant such as vancomycin and heparin, respectively, to a micro arc oxidized surface of a porous additively manufactured implant. The study demonstrates a dynamic implant capable of enhancing bone formation, improving implant bone anchorage and hemocompatibility, and suppressing bacterial growth (Xiu et al. 2016; Zhang et al. 2018). The additively manufactured product post-processing when fused with other technologies produces a hybrid implant. For example, impregnating matrigel into porous Ti-6Al-4V substrate fabricated using EBM for achieving vascularization (Correa et al. 2018) or coating antibiotics such as gentamicin or vancomycin with EPD onto 3D printed porous implant post production to prevent implant infection (Han et al. 2017). Designing implants with internal reservoirs and microchannels for drug delivery are being investigated for reducing biomedical device-associated infections. For preventing implant-associated infection, Ti-6Al-4V hollow cubes fabricated using SLM, filled with vancomycin (an antibiotic), and sealed with Polyether sulfone membrane are being explored (Bezuidenhout et al. 2015). In a similar approach, Cox et al. designed a Ti-6Al-4V hip implant fabricated using SLM with an inbuilt reservoir to elude gentamycin (antibiotic) loaded in calcium phosphate cement (Cox et al. 2016). In addition to fabricating microchannels and internal reservoirs of drug eluting implants, drug release kinetics also needs attention to understand the drug release profile mechanism. This tactic can be further explored in other implants to elute antibiotics post-implantation. Apart from post-processing step to equip the final product with antibacterial and microbial agents, the product itself can be made of metals (such as silver, copper, and zinc) which possess biocidal properties (Turner et al. 2020). Scientists have also explored in situ functionalization while manufacturing, by adding fillers to the main material. Of all the elements, copper has been the most exploited filler due to its biocidal property and affordability when compared to noble metals. The addition of copper in Ti-6Al-4V has shown a significant reduction in growth cultures of bacteria E. coli and S. aureus (Krakhmalev et al. 2017). Innate alloys which possess antibacterial property without additional functionalization are Copper-based alloys. Cobalt-chromium alloys have been used extensively for denture applications. With oral cavity being the hub of bacterial species, an implant with antibacterial properties is highly sought after. The addition of copper to the CoCr alloys equips the alloy with antibacterial activity. CoCrCu and CoCrWCu alloy possessing antimicrobial activity using SLM have been fabricated and tested to combat bacterial infections that take place after implantation. When compared with CoCr alloys without copper, copper containing alloys significantly inhibit biofilm formation and provide an antimicrobial environment (Lu et al. 2018; Ren et al. 2016).

4.2 Additive Manufacturing to Combat COVID-19

In the present scenario, bone implants can also be functionalized with the drug dexamethasone. Dexamethasone, a steroid is one of the triad chemicals required to induce osteogenesis in pluripotent stem cells apart from ascorbic acid and β-glycerol phosphate. Apart from being an inducer of bone formation, clinical trials of dexamethasone have been shown to reduce the mortality rate by one-third in patients infected with COVID-19 (Johnson and Vinetz 2020). Researchers in past, have successfully immobilized dexamethasone on titanium substrate and sustainably using zeolitic imidazolate framework-8 (Johnson and Vinetz 2020) and by self-assembly of Graphene oxide coated with titanate nanostructure to deliver dexamethasone (Ren et al. 2017). EPD, PEO, and other techniques can be explored for deposition of dexamethasone and other COVID combating drugs. Patients undergoing implant surgeries in the current scenario can opt for surface functionalized implant with COVID drugs/vaccines, as a precautionary measure and to boost the immune system. The polymer additive manufacturing industry, however, has been blooming more than metallic additive manufacturing.

5 3D Bioprinted Lung Models

The Severe Acute Respiratory Syndrome Coronavirus 2 (SARS-CoV-2) is a rapidly replicating pathogenic virus that causes severe pneumonia and has transformed into a pandemic resulting in an increased burden on the global healthcare system. The SARS-CoV-2 invades lung cells by binding its spike protein (S), termed peplomers, with a human host cell receptor, angiotensin-converting enzyme II (ACE2) as shown in Fig. 6 (Shyam and Ramalingam n.d.; Hoffmann et al. 2020; VIT team aim to develop aptamer-based handheld biosensors for COVID-19 diagnosis—express healthcare n.d.). ACE2 is a dipeptidyl carboxydipeptidase highly expressed on the luminal surface of epithelial cells which are primarily found in extrapulmonary tissues, for example, heart, kidney, and small intestine (Zhang et al. 2020). In research literature, the presence of ACE2 receptors in the epidermis, alveoli, mucous membrane of eyes, nose, and mouth have been observed in studies conducted during the outbreak of the 2003 SARS caused by the SARS-CoV-1 (Hamming et al. 2004). Overexpression of ACE2 in various in vivo models of human, pig, and civet highlights its roles in the transmission and replication of the virus (Zhou et al., 2020).

Currently, drugs or vaccines that can be used against COVID-19 are at various developmental stages and readers can refer to these sources for further information. (Folegatti et al., 2020; Zhu et al., 2020). Figure 7 highlights the state-of-the-art in global vaccine research and their progress in clinical trials (Zhu et al. 2020). The testing of efficacy and effectiveness of novel drugs and vaccines are carried out on animal models, however, due to physiological differences between animals and humans results in variability. Conversely, It is challenging to study the effects of novel drugs and vaccines at a cellular and molecular level within the human body due to limited access to the lung, where the disease affects the most. Dynamic lung microenvironments which can be recreated through leveraging microfluidic systems with 3D bioprinting are emerging as potential surrogate model systems for the human lungs. These platforms can aid in the study of the efficacy of novel.

Drugs, vaccines, and therapeutic regimes to combat the COVID-19. In this section, the authors provide an overview of advancements in the generation of physiologically similar models of lung architecture and the role of 3DP techniques in the creation of such models and their limitations.

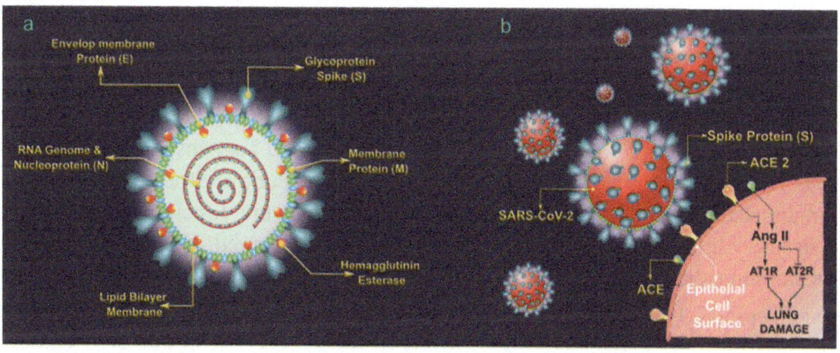

Fig. 6 **a** Schematic representation of the structure of the coronavirus. The "club-like" structures on the envelope of the virus are called peplomers that attach themselves to the ACE2 receptors on the surface of human cells (Shyam and Ramalingam n.d.). **b** Schematic diagram of the Angiotensin system in acute lung failure and proposed SARS-CoV-2 action. Adapted from (VIT team aim to develop aptamer-based handheld biosensors for COVID-19 diagnosis—express healthcare n.d.)

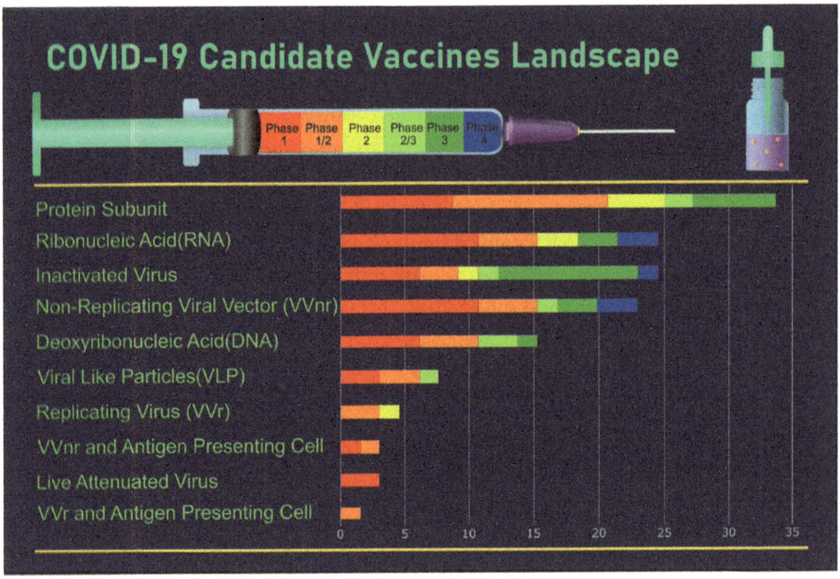

Fig. 7 The development of vaccine against COVID-19. The infographic highlights the types of vaccine and their respective stages of clinical trials (WHO, no date)

5.1 3D in Vitro Lung Tissue Models

The SARS-CoV-2 primarily affects the respiratory system through the transmission of the virus via the ACE2 receptor. ACE2 is expressed in certain organs, e.g., respiratory system, cardiovascular system, renal, and reproductive system. It is responsible for regulating various physiological functions like cell proliferation, hypertrophy, inflammatory response, fluid balance, and blood pressure (Zhang et al. 2020) suggesting that the affinity of the spike protein on the SARS-CoV-2 is to attach itself to the ACE2 receptor and aids in the spread of this infection throughout the body. However, a thorough understanding of the pathogen and host-environment interactions is crucial in the development of effective therapeutic regimes. 2D models of human tissue cannot accurately mimic the in vivo structure, function, and response to drugs, while animal models, which have primarily served as an ideal platform for studying these interactions and for evaluating the efficacy of novel drugs and vaccines, pose a challenge where inter-species differences can hinder translation to clinical trials (Knowlton and Tasoglu 2016; Jimenez-Valdes et al. 2020). Therefore, the need to develop and engineer a complex 3D microenvironment that can emulate human relevant lung environments to study these interactions at a cellular and molecular level is dire. The advancements in tissue engineering have led to the development of 3D models that can aid in a deeper understanding of lung development and pathologies. When combined with microfluidics, perfusable systems which mimic vasculature can be achieved. Additionally, these models offer the ability to mimic both healthy and diseased conditions ex vivo. Figure 8 highlights some of the various models that can be created with varying methods to study pathogen-cell, tissue, or organ interactions at the cellular and molecular levels. Although these systems offer several advantages, there are a few drawbacks associated with each model and these have been summarized in Table 1.

While these models have been generated to replicate the microenvironment of lung tissue for studying various pathological and clinical conditions, to date, no model has been used for the study of the transmission of the SARS-CoV-2, its cellular and molecular interactions, or testing of drugs and vaccines against SARS-CoV-2. This creates a plethora of opportunities in the development of meticulous models that can mimic the pathogen-cell interaction and improve the efficient testing of novel drugs and vaccines.

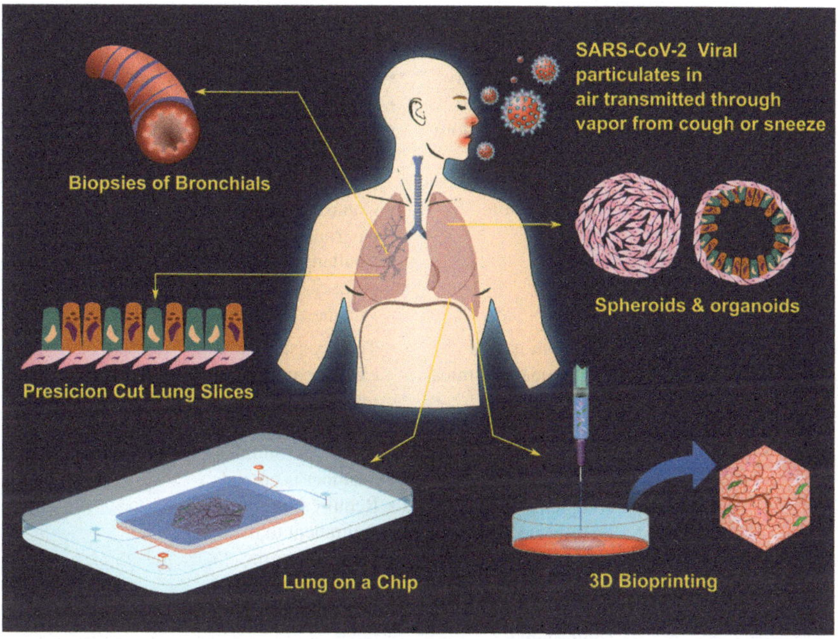

Fig. 8 Complex 3D culture models of human lung for potential COVID-19 studies. Red areas around the nose and mouth indicate high expression of ACE 2 receptors

Table 1 Advantages and disadvantages of the different 3D lung models

Model	Advantages	Disadvantages	Ref.
Bronchial biopsy	• Minimally invasive procedure to obtain samples • Donors can be patients with either lung disorders or healthy lungs • Maintain cellular architecture in 3D environment	• Ex-vivo preservation is challenging • Scarcity of donors • Small quantity of biopsy	Nicholas et al. (2015)
Precision Cut Lung slices	• Disease modelling of lung tissue • Retention of cellular architecture and biological processes	• Small sample size and difficulty in maintaining in Ex-vivo conditions	Alsafadi et al. (2020)
Spheroids & Organoids	• Long term viability can be maintained • Cell-cell interaction can be replicated • Ability to mimic basic organ function	• Lack of vascularisation • Limited mechanical properties that replicate organ function • Control over environment is difficult • Challenging to study efficacy of drugs and vaccines	Lewis et al. (2018)
Lung-on-chip (LOC)	• Ability to mimic mechanical, chemical and biological properties associated with organ function • Ability to replicate complex tissue architecture, especially the air-liquid interface (ALI) • Highly tuned and controllable environment • Can be designed to incorporate vasculature structure • Can be multiplexed with other organ on chip systems to study drug and vaccine efficacy (Human on Chip)	• Time consuming and can be expensive to manufacture • Choice of material to act as microfluidic chamber is crucial • Requires additional equipment to maintain control over the microenvironment • Inability to use all cell types	Benam et al. (2016)

(continued)

Table 1 (continued)

Model	Advantages	Disadvantages	Ref.
3D Bioprinting	• Ability to print complex, controlled, and highly defined architectures • Various materials can be used to replicate tissue extracellular matrix (ECM) and cell types • Endothelial cells can be used to initiate vascularisation with the construct • Maintain long term viability	• Requirement of complex technologies • Tissue specific geometry for architecture is complex and challenging	Grigoryan et al. (2019)

5.2 3D Bioprinting for the Development of 3D Lung Tissue Models

3D bioprinting is a form of additive manufacturing process that involves the printing of tissue-like structures that imitate natural tissues using bioinks, which is a combination of cells and biomaterials. 3D bioprinting can be used to generate complex tissue-like architecture and structures where cellular viability and function are preserved in printed constructs and acts as substitute engineered tissue constructs (Shyam and Ramalingam n.d.; Pati et al. 2016). Despite the various advantages, this technique offers minimal studies on lung organotypic models. Recently, Grigoryan et al. have demonstrated the use of stereolithography to generate a model of alveolar morphology with entangled vascular networks (Grigoryan et al. 2019). Stereolithography is a form of 3DP where, based on a digital pattern in 3D space, a photon source is focused on a photosensitive liquid to induce polymerization and transforming it into a polymerized solid (Sun et al. 2005). The model by Grigoryan et al. was printed with poly (ethylene glycol) diacrylate (PEGDA), a water-soluble photosensitive polymer (Husar et al. 2014), and perfused with red blood cells (RBC) to mimic the alveolar sac (Grigoryan et al. 2019). Though Grigoryan et al. did not conduct any studies related to the performance of drugs on this model, the ability to construct complex tissue structures that physiologically mimic in vivo conditions presents an opportunity in the development of models. Such models can be used to study the interaction of pathogen at cellular and molecular level and study the efficacy of novel drugs. On the other hand, developments in the implementation of 3DP for the generation of microfluidic systems and disease modeling are also emerging. There can be two approaches in the utilization of 3DP. In the first approach, the lung architecture can be mimicked on a microfluidic chamber as shown in Fig. 9a. For example, Benam et al. fabricated a device composed of two parallel microchannels separated by a rigid porous

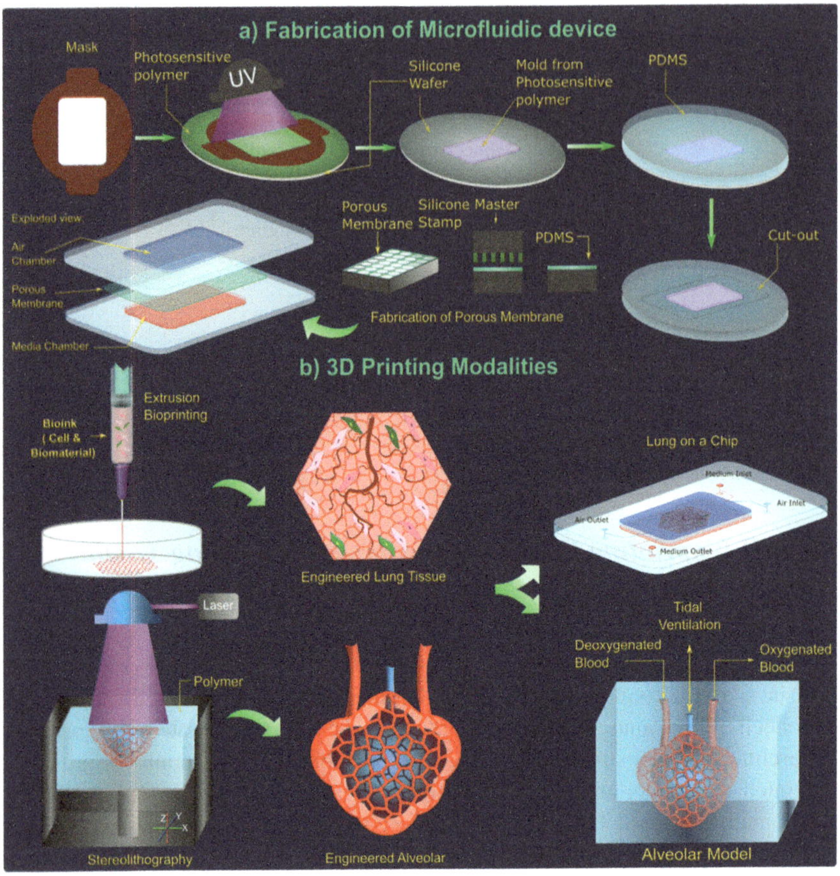

Fig. 9 Schematic representation of the fabrication of a conventional lung on chip and the incorporation of 3D bioprinting to generate next generation lung on chip models for COVID-19 testing

membrane. In one channel, primary human airway bronchiolar epithelial cells were cultured, while the opposing channel consists of lung microvascular endothelial cells interfaced by the porous membrane creating an epithelium–endothelium interface. Neutrophil enriched culture medium flows through the microvascular channel. Such small models on a chip present a new tool to model complex and dynamic inflammatory responses of lung in vitro (Benam et al. 2016). One of the major challenges with such models is that the substrate used to recreate the porous membrane is made from synthetic materials, which is rigid, and does not replicate the physiological features inherent in the lung. A possible solution for this challenge could be 3D bioprinting the membrane using hydrogels. This can be achieved through the development of bioinks composed of lung-like ECM. The advantage of encapsulating a lung equivalent tissue on a cellular scale is the ability to study drug-cell-microenvironment interactions and the ability to precisely control dose. In a second approach, 3D bioprinting can be employed to print complex architectures like the alveolar sac. This is possible due to the current technological means to capture its intricate physiological organ and tissue architecture via biomedical imaging such as CT Scan and MRI. These methods can be employed to capture details such as vasculature networks. The images can be used to generate 3D models which can be printed to human relevant size to study lung diseases. Most importantly, 3D bioprinting can provide high resolution where due to its ability to print patient-specific engineered constructs, the regeneration of damaged tissue in patients who have suffered from the detrimental effects of COVID-19 can live prolonged life through the implantation of patient-specific artificial constructs.

Figure 9 highlights the possible process in the implementation of microfluidics and 3D bioprinting in the creation of platforms for testing new or repurposed drugs and vaccines against COVID-19. One of the major challenges of the implementation of 3D bioprinting is the requirement of complex equipment. Additionally, the choice of biomaterials and cells that can successfully mimic lung extracellular matrix (ECM) and function are crucial considerations conducive for generating engineered tissues capable of accurately mimicking native lung tissue.

5.3 Challenges of 3D Bioprinting of Lung Tissue

The capability to mimic physiological cellular interactions through leveraging 3D bioprinting have been demonstrated in several publications for several organs (Tasnim et al. 2018). However, these devices remain a niche tool in research and there are several challenges in the deployment of 3D bioprinting in the assessment of novel drugs. One of the major challenges is that 3D bioprinting does not provide extremely high resolution, deemed necessary to recapitulate organ function at this scale. Another challenge is the vascularization of the engineered tissue. Within the human body, the vascular network provides oxygen and nutrients to cells and is responsible for the removal of metabolic waste. Additionally, it is also responsible for the delivery of drug and therapeutic modalities to diseased organs. Hence, designing an efficient vasculature network that can ensure the survivability of the cells and act as an efficient model to study the lung ex vivo is crucial. Therefore, the generation of an efficient model that most accurately mimics the native lung architecture including an efficient vascular network is the need of the hour, specifically in the current pandemic situation where lung models can provide a deeper understanding of SARS-CoV-2 ingression, thereby enabling the designing of even more efficacious drugs and vaccines.

6 3D Printed Drug Delivery Systems for Covid-19

Personalized or customizable medicine is the new buzz word in the pharmaceutical or health care sector, in which single medicine, or their combinations or their dose are customized to suit individual needs (Trenfield et al. 2019). 3DP is being explored as a potential technology to realize the above-mentioned needs. Also, 3DP-based technologies could be utilized to make bespoke drug delivery systems of any shape and size in a layer-by-layer fashion. Thus, 3DP technologies could be valuable for such customization, which is not achievable using conventional mass manufacturing processes. Moreover, through such customizable medicines, patient compliance and tailored drug release profiles could be achieved (Trenfield et al. 2019; Norman et al. 2017).

6.1 3D Printed Bespoke Drug Delivery Systems

3DP is used to fabricate complex and custom-made drug formulations which might not be possible using the conventional manufacturing processes. 3DP was reported to implement a wide variety of formulations with tailored release profiles and designs, ranging from controlled release systems, fast dissolving tablets, and multi-drug combinatory systems (Norman et al. 2017). The printed tablets are called as printlets, which are designed into complex shapes as well as by varying the infill percentage, and polymer inclusion to modify release kinetics (Sadia et al. 2018).

In one study, the printlets were fabricated into cylindrical and gyroid structures using SLS and was found to achieve customizable release characteristics based on the geometry selected (Fina et al. 2018) Fig. 10a, b. Theophylline-loaded printlets with a radiator-like design has been fabricated using FDM printing (Isreb et al. 2019) Fig. 10f. In order to vary the drug release kinetics from the printed structure, each dosage form had connected paralleled plates with inter-plate spacing of either 0.5, 1, 1.5 or 2 mm (Isreb et al. 2019). Printlets with lower infill percentages have faster drug release kinetics when compared to the higher infill ones (Goyanes et al. 2014). In another study by Chai et al., lower infill percentage was used for gastro-retentive tablets by tapping its buoyancy property (Chai et al. 2017). Highly porous and fast dissolving printlets were fabricated using certain 3DP processes (such as SLS and binder jetting) (Alhnan et al. 2016). Fina et al. demonstrated the fabrication of or dispersible printlets for the first time using SLS and by simply changing the laser speed at which the powder particles were sintered (Fina et al. 2018). Spritam (levetiracetam), is the first FDA approved 3D printed drug, orally administered for the treatment of seizures in adults and children with epilepsy (Prasad and Smyth 2016).

Another research area in which 3DP has been widely researched is the production of multi-drug combinations (or polypills). Pereira et al. fabricated a four-drug cardio-vascular polypill (Pereira et al. 2019) (Fig. 10d). In another study, 3D printed cylindrical and ring-shaped polypills were fabricated to encapsulate six different drugs (paracetamol, naproxen, caffeine, prednisolone, aspirin, and chloramphenicol) to improve the patients' medication adherence (Robles-Martinez et al. 2019). In recent research, Awad et al. fabricated 3D printed pellets (mini printlets) containing two

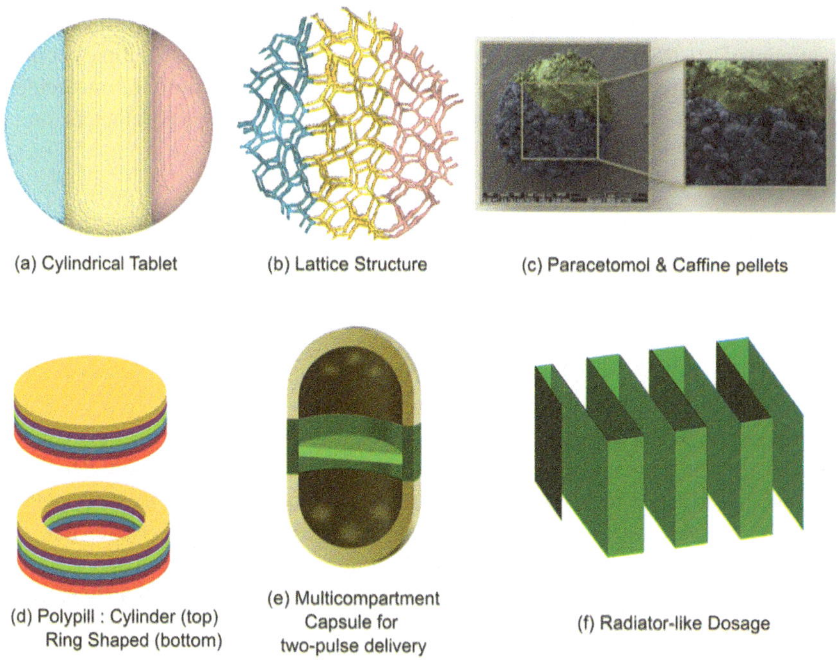

Fig. 10 3DP of pills and tablets using either SLS, SLA or FDM methods. **a** Cylindrical form **b** lattice structure with different compositions/dosage/time release (Fina et al. 2018) **c** combination of two drugs (Awad et al. 2019), **d** polypill design and structure (Robles-Martinez et al. 2019), **e** multicompartment pill (Maroni et al. 2017), **f** "radiator" design (Isreb et al. 2019)

drugs which are spatially separated and thereby engineered for varying drug release profile (Awad et al. 2019; Maroni et al. 2017) (Fig. 10e).

6.2 3D Printed Drug Delivery Systems for COVID-19

3DP has been used to fabricate drug-containing nose masks custom-made for the patients for the treatment of acne (Goyanes et al. 2016). In the study, patient's nose region was scanned using a 3D scanner, followed by developing a 3D model which could create a mask that was personalized to the patient. Muwaffak et al., demonstrated 3D printed custom-made wound dressings which has zinc, silver, and copper as their antimicrobial agents in the shape of a nose and an ear (Muwaffak et al. 2017). This helps in retaining the dressings in the wound region when compared to their analogous flat dressings. In case of combating COVID-19, a similar approach can be utilized to fabricate antiviral drugs/molecules containing face masks.

Also, there are reports that the combination of Lopinavir (LPV) and Ritonavir (RTV) with Ribavirin is an effective antiviral agent against SARS. The combination therapy consisting of interferon-β, LPV/RTV, and Ribavirin is effective against the treatment of MERS-CoV disease (Chauhan et al. 2020). This kind of complex combination therapeutic formulations could be 3D printed and each antiviral molecules' release profile could be fine-tuned based on the 3D model design parameters.

7 Conclusion and Future Work

In this chapter, the emerging role that the 3DP technologies played during the global lockdown due to COVID-19 pandemic and their implementation in the manufacturing of key PPEs, critical medical devices like ventilator valves, devices for use in testing kits, and innovative methods to reduce contact with surfaces prone to harbor the virus has been discussed. The scope of 3DP biomedical implants, and the importance of post-processing, along with the need for surface modification and drug release kinetics highlights the need to implement this manufacturing method in future research. Of interest is the embedding of drugs or nanoparticles that can combat COVID-19, or similar virus, which requires further research and development towards translational medicine. Secondly, the development of human lung tissue equivalent platforms and the implementation of 3D bioprinting to print functional lung tissues were discussed. These 3D bioprinted lung tissue models could act as a platform for antiviral drug or vaccine screening, as well as to understand their toxicity profiles. While single organ models are unsuccessful in replicating the complexity and integrity associated with functional organs, there is much research attention on the incorporation of vascularization to improve the functioning of the 3D printed organs. Moreover, by incorporating medical imaging modalities, patient derived stem cells and 3DP, personalized medicine can be designed for patients most severely affected by SARS-CoV-2. Additionally, it can aid in the designing of vaccines and drugs for patients for whom commercial vaccine may not necessarily work. As research and development in the generation of improved drugs to combat COVID-19 ensues, 3DP is making a steady way into the printing of medicine termed printlets. Printlets manufactured via 3DP offer the versatility of being custom-designed to suit patients' needs along with incorporating multiple doses and with varying release kinetics, within a singular tablet. 3DP is evolving at a rapid rate, with extensive research funding being deployed in this technology as a result of the benefits this technology had to offer during pandemic scale events as witnessed in the year 2020 due to COVID-19. 3DP offered solutions where conventional manufacturing methods failed to deliver as a result of global lockdown and block in the supply chain. The critical medical equipment or their parts and PPE's could be custom-made, on-demand in areas most affected by an influx of cases of COVID-19 and limited supply of medical care. Towards the future, it is strongly suggested that researchers employ strategies that can enhance the implementation of 3DP in various aspects, as well as educate the regulators and policymakers about its positive impact, which might result in the faster approval of 3D printed medical products. A larger focus towards the adoption of simplified regulations towards commodities manufactured via 3DP, without compromise to quality and safety, is the need of the hour.

Acknowledgement Dr. Arunkumar Palaniappan would like to kindly acknowledge the financial support from the Science and Engineering Re-search Board (SERB), Department of Science and Technology, Government of India through its start-up research grant scheme (SRG/2020/001115).

Dr. Loganathan Rangasamy would like to kindly acknowledge the financial supports from the Department of Bio-technology, Government of India, through Ramalingaswami Re-entry fellowship (BT/RLF/Reentry/44/2018), Science and Engineering Research Board (SERB), New Delhi for a Core Research Grant (CRG) (CRG/2020/001213) and VIT SEED GRANT 2020-21.

References

Alhnan MA, Okwuosa TC, Sadia M, Wan KW, Ahmed W, Arafat B (2016) Emergence of 3D printed dosage forms: opportunities and challenges. Pharm Res. https://doi.org/10.1007/s11095-016-1933-1

Alsafadi HN, Uhl FE, Pineda RH, Bailey KE, Rojas M, Wagner DE, Königshoff M (2020) Applications and approaches for three-dimensional precision-cut lung slices. disease modeling and drug discovery. Am J Respir Cell Mol Biol 62:681–691. https://doi.org/10.1165/rcmb.2019-0276TR

Amin D, Nguyen N, Roser SM, Abramowicz S (2020) 3D printing of face shields during COVID-19 pandemic: a technical note. J Oral Maxillofac Surg 1–4. https://doi.org/10.1016/j.joms.2020.04.040

Awad A, Fina F, Trenfield S, Patel P, Goyanes A, Gaisford S, Basit A (2019) 3D printed pellets (Miniprintlets): a novel, multi-drug, controlled release platform technology. Pharmaceutics 11:148. https://doi.org/10.3390/pharmaceutics11040148

Benam KH, Villenave R, Lucchesi C, Varone A, Hubeau C, Lee HH, Alves SE, Salmon M, Ferrante TC, Weaver JC, Bahinski A, Hamilton GA, Ingber DE (2016) Small airway-on-a-chip enables analysis of human lung inflammation and drug responses in vitro. Nat Methods 13:151–157. https://doi.org/10.1038/nmeth.3697

Bezuidenhout MB, van Staden AD, Oosthuizen GA, Dimitrov DM, Dicks LMT (2015) Delivery of antibiotics from cementless titanium-alloy cubes may be a novel way to control postoperative infections. Biomed Res Int 2015:856859. https://doi.org/10.1155/2015/856859

Chai X, Chai H, Wang X, Yang J, Li J, Zhao Y, Cai W, Tao T, Xiang X (2017) Fused deposition modeling (FDM) 3D printed tablets for intragastric floating delivery of domperidone. Sci Rep. https://doi.org/10.1038/s41598-017-03097-x

Chauhan DS, Prasad R, Srivastava R, Jaggi M, Chauhan SC, Yallapu MM (2020) Comprehensive review on current interventions, diagnostics, and nanotechnology perspectives against SARS-CoV-2. Bioconjug Chem. https://doi.org/10.1021/acs.bioconjchem.0c00323

Chen X, Possel JK, Wacongne C, van Ham AF, Klink PC, Roelfsema PR (2017) 3D printing and modelling of customized implants and surgical guides for non-human primates. J Neurosci Methods 286:38–55. https://doi.org/10.1016/j.jneumeth.2017.05.013

Choudhari CM, Patil VD (2016) Product development and its comparative analysis by SLA, SLS and FDM rapid prototyping processes. IOP Conf Ser Mater Sci Eng 149. https://doi.org/10.1088/1757-899X/149/1/012009

Chudinova EA, Surmeneva MA, Timin AS, Karpov TE, Wittmar A, Ulbricht M, Ivanova A, Loza K, Prymak O, Koptyug A, Epple M, Surmenev RA (2019) Adhesion, proliferation, and osteogenic differentiation of human mesenchymal stem cells on additively manufactured Ti6Al4V alloy scaffolds modified with calcium phosphate nanoparticles. Colloids Surf B Biointerfaces 176:130–139. https://doi.org/10.1016/j.colsurfb.2018.12.047

Correa VL, Garza KM, Murr LE (2018) Vascularization in interconnected 3D printed Ti-6Al-4V foams with hydrogel matrix for biomedical bone replacement implants. Sci China Mater 61:556–578. https://doi.org/10.1007/s40843-017-9091-1

Cox SC, Jamshidi P, Eisenstein NM, Webber MA, Hassanin H, Attallah MM, Shepherd DET, Addison O, Grover LM (2016) Adding functionality with additive manufacturing: fabrication of titanium-based antibiotic eluting implants. Mater Sci Eng C 64:407–415. https://doi.org/10.1016/j.msec.2016.04.006

Ćwikła G, Grabowik C, Kalinowski K, Paprocka I, Ociepka P (2017) The influence of printing parameters on selected mechanical properties of FDM/FFF 3D-printed parts. IOP Conf Ser Mater Sci Eng 227. https://doi.org/10.1088/1757-899X/227/1/012033

Draft landscape of COVID-19 candidate vaccines (n.d.). https://www.who.int/publications/m/item/draft-landscape-of-covid-19-candidate-vaccines. Accessed 4 Oct 2020

Dzian A, Živčák J, Penciak R, Hudák R (2018) Implantation of a 3D-printed titanium sternum in a patient with a sternal tumor. World J Surg Oncol 16. https://doi.org/10.1186/s12957-018-1315-8

Fina F, Goyanes A, Madla CM, Awad A, Trenfield SJ, Kuek JM, Patel P, Gaisford S, Basit AW (2018) 3D printing of drug-loaded gyroid lattices using selective laser sintering. Int J Pharm 547:44–52. https://doi.org/10.1016/j.ijpharm.2018.05.044

Folegatti PM, Ewer KJ, Aley PK, Angus B, Becker S, Belij-Rammerstorfer S, Bellamy D, Bibi S, Bittaye M, Clutterbuck EA, Dold C, Faust SN, Finn A, Flaxman AL, Hallis B, Heath P, Jenkin D, Lazarus R, Makinson R, Minassian AM, Pollock KM, Ramasamy M, Robinson H, Snape M, Tarrant R, Voysey M, Green C, Douglas AD, Hill AVS, Lambe T, Gilbert SC, Pollard AJ, Aboagye J, Adams K, Ali A, Allen E, Allison JL, Anslow R, Arbe-Barnes EH, Babbage G, Baillie K, Baker M, Baker N, Baker P, Baleanu I, Ballaminut J, Barnes E, Barrett J, Bates L, Batten A, Beadon K, Beckley R, Berrie E, Berry L, Beveridge A, Bewley KR, Bijker EM, Bingham T, Blackwell L, Blundell CL, Bolam E, Boland E, Borthwick N, Bower T, Boyd A, Brenner T, Bright PD, Brown-O'Sullivan C, Brunt E, Burbage J, Burge S, Buttigieg KR, Byard N, Cabera Puig I, Calvert A, Camara S, Cao M, Cappuccini F, Carr M, Carroll MW, Carter V, Cathie K, Challis RJ, Charlton S, Chelysheva I, Cho JS, Cicconi P, Cifuentes L, Clark H, Clark E, Cole T, Colin-Jones R, Conlon CP, Cook A, Coombes NS, Cooper R, Cosgrove CA, Coy K, Crocker WEM, Cunningham CJ, Damratoski BE, Dando L, Datoo MS, Davies H, De Graaf H, Demissie T, Di Maso C, Dietrich I, Dong T, Donnellan FR, Douglas N, Downing C, Drake J, Drake-Brockman R, Drury RE, Dunachie SJ, Edwards NJ, Edwards FDL, Edwards CJ, Elias SC, Elmore MJ, Emary KRW, English MR, Fagerbrink S, Felle S, Feng S, Field S, Fixmer C, Fletcher C, Ford KJ, Fowler J, Fox P, Francis E, Frater J, Furze J, Fuskova M, Galiza E, Gbesemete D, Gilbride C, Godwin K, Gorini G, Goulston L, Grabau C, Gracie L, Gray Z, Guthrie LB, Hackett M, Halwe S, Hamilton E, Hamlyn J, Hanumunthadu B, Harding I, Harris SA, Harris A, Harrison D, Harrison C, Hart TC, Haskell L, Hawkins S, Head I, Henry JA, Hill J, Hodgson SHC, Hou MM, Howe E, Howell N, Hutlin C, Ikram S, Isitt C, Iveson P, Jackson S, Jackson F, James SW, Jenkins M, Jones E, Jones K, Jones CE, Jones B, Kailath R, Karampatsas K, Keen J, Kelly S, Kelly D, Kerr D, Kerridge S, Khan L, Khan U, Killen A, Kinch J, King TB, King L, King J, Kingham-Page L, Klenerman P, Knapper F, Knight JC, Knott D, Koleva S, Kupke A, Larkworthy CW, Larwood JPJ, Laskey A, Lawrie AM, Lee A, Ngan Lee KY, Lees EA, Legge H, Lelliott A, Lemm NM, Lias AM, Linder A, Lipworth S, Liu X, Liu S, Lopez Ramon R, Lwin M, Mabesa F, Madhavan M, Mallett G, Mansatta K, Marcal I, Marinou S, Marlow E, Marshall JL, Martin J, McEwan J, McInroy L, Meddaugh G, Mentzer AJ, Mirtorabi N, Moore M, Moran E, Morey E, Morgan V, Morris SJ, Morrison H, Morshead G, Morter R, Mujadidi YF, Muller J, Munera-Huertas T, Munro C, Munro A, Murphy S, Munster VJ, Mweu P, Noé A, Nugent FL, Nuthall E, O'Brien K, O'Connor D, Oguti B, Oliver JL, Oliveira C, O'Reilly PJ, Osborn M, Osborne P, Owen C, Owens D, Owino N, Pacurar M, Parker K, Parracho H, Patrick-Smith M, Payne V, Pearce J, Peng Y, Peralta Alvarez MP, Perring J, Pfafferott K, Pipini D, Plested E, Pluess-Hall H, Pollock K, Poulton I, Presland L, Provstgaard-Morys S, Pulido D, Radia K, Ramos Lopez F, Rand J, Ratcliffe H, Rawlinson T,

Rhead S, Riddell A, Ritchie AJ, Roberts H, Robson J, Roche S, Rohde C, Rollier CS, Romani R, Rudiansyah I, Saich S, Sajjad S, Salvador S, Sanchez Riera L, Sanders H, Sanders K, Sapaun S, Sayce C, Schofield E, Screaton G, Selby B, Semple C, Sharpe HR, Shaik I, Shea A, Shelton H, Silk S, Silva-Reyes L, Skelly DT, Smee H, Smith CC, Smith DJ, Song R, Spencer AJ, Stafford E, Steele A, Stefanova E, Stockdale L, Szigeti A, Tahiri-Alaoui A, Tait M, Talbot H, Tanner R, Taylor IJ, Taylor V, Te Water Naude R, Thakur N, Themistocleous Y, Themistocleous A, Thomas M, Thomas TM, Thompson A, Thomson-Hill S, Tomlins J, Tonks S, Towner J, Tran N, Tree JA, Truby A, Turkentine K, Turner C, Turner N, Turner S, Tuthill T, Ulaszewska M, Varughese R, Van Doremalen N, Veighey K, Verheul MK, Vichos I, Vitale E, Walker L, Watson MEE, Welham B, Wheat J, White C, White R, Worth AT, Wright D, Wright S, Yao XL, Yau Y (2020) Safety and immunogenicity of the ChAdOx1 nCoV-19 vaccine against SARS-CoV-2: a preliminary report of a phase 1/2, single-blind, randomised controlled trial. Lancet 396:467–478. https://doi.org/10.1016/S0140-6736(20)31604-4

Gallup N, Pringle AM, Oberloier S, Tanikella NG, Pearce JM (2020) Parametric nasopharyngeal swab for sampling COVID-19 and other respiratory viruses: open source design, SLA 3-D printing and UV curing system. HardwareX 8:e00135. https://doi.org/10.1016/j.ohx.2020.e00135

Goyanes A, Buanz ABM, Basit AW, Gaisford S (2014) Fused-filament 3D printing (3DP) for fabrication of tablets. Int J Pharm. https://doi.org/10.1016/j.ijpharm.2014.09.044

Goyanes A, Det-Amornrat U, Wang J, Basit AW, Gaisford S (2016) 3D scanning and 3D printing as innovative technologies for fabricating personalized topical drug delivery systems. J Control Release. https://doi.org/10.1016/j.jconrel.2016.05.034

Grigoryan B, Paulsen SJ, Corbett DC, Sazer DW, Fortin CL, Zaita AJ, Greenfield PT, Calafat NJ, Gounley JP, Ta AH, Johansson F, Randles A, Rosenkrantz JE, Louis-Rosenberg JD, Galie PA, Stevens KR, Miller JS (2019) Multivascular networks and functional intravascular topologies within biocompatible hydrogels. Science 364(80):458–464. http://science.sciencemag.org/

Guan B, Wang H, Xu R, Zheng G, Yang J, Liu Z, Cao M, Wu M, Song J, Li N, Li T, Cai Q, Yang X, Li Y, Zhang X (2016) Establishing antibacterial multilayer films on the surface of direct metal laser sintered titanium primed with phase-transited lysozyme. Sci Rep 6:1–12. https://doi.org/10.1038/srep36408

Guo Y, Ren L, Xie K, Wang L, Yu B, Jiang W, Zhao Y, Hao Y (2020) Functionalized TiCu/Ti–Cu–N–Coated 3D-printed porous Ti6Al4V scaffold promotes bone regeneration through BMSC recruitment. Adv Mater Interfaces 7:1901632. https://doi.org/10.1002/admi.201901632

Hamming I, Timens W, Bulthuis MLC, Lely AT, Navis GJ, van Goor H (2004) Tissue distribution of ACE2 protein, the functional receptor for SARS coronavirus. A first step in understanding SARS pathogenesis. J Pathol 203:631–637. https://doi.org/10.1002/path.1570

Han C, Yao Y, Cheng X, Luo J, Luo P, Wang Q, Yang F, Wei Q, Zhang Z (2017) Electrophoretic deposition of gentamicin-loaded silk fibroin coatings on 3D-printed porous cobalt-chromium-molybdenum bone substitutes to prevent orthopedic implant infections. Biomacromol 18:3776–3787. https://doi.org/10.1021/acs.biomac.7b01091

Hoffmann M, Kleine-Weber H, Schroeder S, Krüger N, Herrler T, Erichsen S, Schiergens TS, Herrler G, Wu NH, Nitsche A, Müller MA, Drosten C, Pöhlmann S (2020) SARS-CoV-2 Cell Entry Depends on ACE2 and TMPRSS2 and Is blocked by a clinically proven protease inhibitor. Cell 181:271–280.e8. https://doi.org/10.1016/j.cell.2020.02.052

Husar B, Hatzenbichler M, Mironov V, Liska R, Stampfl J, Ovsianikov A (2014) Photopolymerization-based additive manufacturing for the development of 3D porous scaffolds. Biomater Bone Regen Nov Tech Appl (Elsevier Ltd) 149–201. https://doi.org/10.1533/9780857098104.2.149

Ishack S, Lipner SR (2020) Applications of 3D printing technology to address COVID-19—related supply shortages. Am J Med 2020:2019–2021. https://doi.org/10.1016/j.amjmed.2020.04.002

Isreb A, Baj K, Wojsz M, Isreb M, Peak M, Alhnan MA (2019) 3D printed oral theophylline doses with innovative 'radiator-like' design: impact of polyethylene oxide (PEO) molecular weight. Int J Pharm 564:98–105. https://doi.org/10.1016/j.ijpharm.2019.04.017

Jardini AL, Larosa MA, Macedo MF, Bernardes LF, Lambert CS, Zavaglia CAC, Filho RM, Calderoni DR, Ghizoni E, Kharmandayan P (2016) Improvement in cranioplasty: advanced prosthesis biomanufacturing. In: Procedia CIRP. Elsevier B.V., pp 203–208. https://doi.org/10.1016/j.procir.2015.11.017

Jia Z, Zhou W, Yan J, Xiong P, Guo H, Cheng Y, Zheng Y (2018) constructing multilayer silk protein/nanosilver biofunctionalized hierarchically structured 3D printed Ti6Al4V scaffold for repair of infective bone defects. ACS Biomater Sci Eng 5:244–261. https://doi.org/10.1021/acsbiomaterials.8b00857

Jiménez M, Romero L, Domínguez I, Complexity ME (2019) Additive manufacturing technologies: an overview about 3D printing methods and future prospects, Hindawi.Com. (n.d.)

Jiménez M, Romero L, Domínguez IA, Espinosa MDM, Domínguez M (2019) Additive manufacturing technologies: an overview about 3D printing methods and future prospects. Complexity. https://doi.org/10.1155/2019/9656938

Jimenez-Valdes RJ, Can UI, Niemeyer BF, Benam KH (2020) Where we stand: lung organotypic living systems that emulate human-relevant host–environment/pathogen interactions. Front Bioeng Biotechnol 8:989. https://doi.org/10.3389/fbioe.2020.00989

Johnson RM, Vinetz JM (2020) Dexamethasone in the management of COVID-19. BMJ 370. https://doi.org/10.1136/bmj.m2648

Karunakaran KP, Suryakumar S, Pushpa V, Akula S (2010) Low cost integration of additive and subtractive processes for hybrid layered manufacturing. Robot Comput Integr Manuf 26:490–499. https://doi.org/10.1016/j.rcim.2010.03.008

Knowlton S, Tasoglu S (2016) A bioprinted liver-on-a-chip for drug screening applications. Trends Biotechnol 34:681–682. https://doi.org/10.1016/j.tibtech.2016.05.014

Krakhmalev P, Yadroitsev I, Yadroitsava I, de Smidt O (2017) Functionalization of biomedical Ti6-Al-4V via in situ alloying by Cu during laser powder bed fusion manufacturing. Materials (Basel) 10. https://doi.org/10.3390/ma10101154

Lewis KJR, Hall JK, Kiyotake EA, Christensen T, Balasubramaniam V, Anseth KS (2018) Epithelial-mesenchymal crosstalk influences cellular behavior in a 3D alveolus-fibroblast model system. Biomaterials 155:124–134. https://doi.org/10.1016/j.biomaterials.2017.11.008

Li J, Mutreja I, Hooper GJ, Clinch K, Lim K, Evans G, Woodfield TF (2020) Combined infection control and enhanced osteogenic differentiation capacity on additive manufactured Ti-6Al-4V are mediated via titania nanotube delivery of novel biofilm inhibitors. Adv Mater Interfaces 7:1901963. https://doi.org/10.1002/admi.201901963

Lu Y, Ren L, Wu S, Yang C, Lin W, Xiao S, Yang Y, Yang K, Lin J (2018) CoCrWCu alloy with antibacterial activity fabricated by selective laser melting: densification, mechanical properties and microstructural analysis. Powder Technol 325:289–300. https://doi.org/10.1016/j.powtec.2017.11.018

Malyala SK, Kumar R, Alwala AM (2017) A 3D-printed osseointegrated combined jaw and dental implant prosthesis—a case study. Rapid Prototyp J 23:1164–1169. https://doi.org/10.1108/RPJ-10-2016-0166

Mangano C, Bianchi A, Mangano FG, Dana J, Colombo M, Solop I, Admakin O (2020) Custom-made 3D printed subperiosteal titanium implants for the prosthetic restoration of the atrophic posterior mandible of elderly patients: a case series. 3D Print Med 6:1. https://doi.org/10.1186/s41205-019-0055-x

Maroni A, Melocchi A, Parietti F, Foppoli A, Zema L, Gazzaniga A (2017) 3D printed multi-compartment capsular devices for two-pulse oral drug delivery. J Control Release 268:10–18. https://doi.org/10.1016/j.jconrel.2017.10.008

Merklein M, Junker D, Schaub A, Neubauer F (2016) Hybrid additive manufacturing technologies—an analysis regarding potentials and applications. Phys Procedia 83:549–559. https://doi.org/10.1016/j.phpro.2016.08.057

Muwaffak Z, Goyanes A, Clark V, Basit AW, Hilton ST, Gaisford S (2017) Patient-specific 3D scanned and 3D printed antimicrobial polycaprolactone wound dressings. Int J Pharm. https://doi.org/10.1016/j.ijpharm.2017.04.077

Nicholas B, Staples KJ, Moese S, Meldrum E, Ward J, Dennison P, Havelock T, Hinks TSC, Amer K, Woo E, Chamberlain M, Singh N, North M, Pink S, Wilkinson TMA, Djukanović R (2015) A novel lung explant model for the ex vivo study of efficacy and mechanisms of anti-influenza drugs. J Immunol 194:6144–6154. https://doi.org/10.4049/jimmunol.1402283

Norman J, Madurawe RD, Moore CMV, Khan MA, Khairuzzaman A (2017) A new chapter in pharmaceutical manufacturing: 3D-printed drug products. Adv Drug Deliv Rev. https://doi.org/10.1016/j.addr.2016.03.001

Park EK, Lim JY, Yun IS, Kim JS, Woo SH, Kim DS, Shim KW (2016) Cranioplasty enhanced by three-dimensional printing: custom-made three-dimensional-printed titanium implants for skull defects. J Craniofac Surg 27:943–949. https://doi.org/10.1097/SCS.0000000000002656

Pati F, Gantelius J, Svahn HA (2016) 3D bioprinting of tissue/organ models. Angew Chem Int Ed 55:4650–4665. https://doi.org/10.1002/anie.201505062

Pereira BC, Isreb A, Forbes RT, Dores F, Habashy R, Petit JB, Alhnan MA, Oga EF (2019) 'Temporary Plasticiser': A novel solution to fabricate 3D printed patient-centred cardiovascular 'Polypill' architectures. Eur J Pharm Biopharm. https://doi.org/10.1016/j.ejpb.2018.12.009

Phan K, Sgro A, Maharaj MM, D'Urso P, Mobbs RJ (2016) Application of a 3D custom printed patient specific spinal implant for C1/2 arthrodesis. J Spine Surg 2:314–318. https://doi.org/10.21037/jss.2016.12.06

Prasad LK, Smyth H (2016) 3D Printing technologies for drug delivery: a review. Drug Dev Ind Pharm. https://doi.org/10.3109/03639045.2015.1120743

Ren L, Memarzadeh K, Zhang S, Sun Z, Yang C, Ren G, Allaker RP, Yang K (2016) A novel coping metal material CoCrCu alloy fabricated by selective laser melting with antimicrobial and antibiofilm properties. Mater Sci Eng C 67:461–467. https://doi.org/10.1016/j.msec.2016.05.069

Ren N, Li J, Qiu J, Yan M, Liu H, Ji D, Huang J, Yu J, Liu H (2017) Growth and accelerated differentiation of mesenchymal stem cells on graphene-oxide-coated titanate with dexamethasone on surface of titanium implants. Dent Mater 33:525–535. https://doi.org/10.1016/j.dental.2017.03.001

Robles-Martinez P, Xu X, Trenfield SJ, Awad A, Goyanes A, Telford R, Basit AW, Gaisford S (2019) 3D printing of a multi-layered polypill containing six drugs using a novel stereolithographic method. Pharmaceutics 11. https://doi.org/10.3390/pharmaceutics11060274

Rodriguez-Palomo A, Monopoli D, Afonso H, Izquierdo-Barba I, Vallet-Regí M (2016) Surface zwitterionization of customized 3D Ti6Al4V scaffolds: a promising alternative to eradicate bone infection. J Mater Chem B 4:4356–4365. https://doi.org/10.1039/c6tb00675b

Rodríguez-Panes A, Claver J, Camacho AM (2018) The influence of manufacturing parameters on the mechanical behaviour of PLA and ABS pieces manufactured by FDM: a comparative analysis. Materials (Basel) 11. https://doi.org/10.3390/ma11081333

Sadia M, Arafat B, Ahmed W, Forbes RT, Alhnan MA (2018) Channelled tablets: an innovative approach to accelerating drug release from 3D printed tablets. J Control Release. https://doi.org/10.1016/j.jconrel.2017.11.022

Shuang F, Hu W, Shao Y, Li H, Zou H (2016) Treatment of intercondylar humeral fractures with 3D-printed osteosynthesis plates. Med (United States) 95. https://doi.org/10.1097/MD.0000000000002461

Shyam R, Ramalingam M (n.d.) 3D printing to play major role in mitigating spread of COVID-19. Say Researchers from VIT—MANUFACTUR3D. https://manufactur3dmag.com/3d-printing-to-play-major-role-in-mitigating-spread-of-covid-19-say-researchers-from-vit/. Accessed 1 Sep 2020

Sun C, Fang N, Wu DM, Zhang X (2005) Projection micro-stereolithography using digital micro-mirror dynamic mask. Sens Actuators A 121:113–120. https://doi.org/10.1016/j.sna.2004.12.011

Surmeneva M, Lapanje A, Chudinova E, Ivanova A, Koptyug A, Loza K, Prymak O, Epple M, Ennen-Roth F, Ulbricht M, Rijavec T, Surmenev R (2019) Decreased bacterial colonization of additively manufactured Ti6Al4V metallic scaffolds with immobilized silver and calcium phosphate nanoparticles. Appl Surf Sci 480:822–829. https://doi.org/10.1016/j.apsusc.2019.03.003

Swennen GRJ, Pottel L, Haers PE (2020) Custom-made 3D-printed face masks in case of pandemic crisis situations with a lack of commercially available FFP2/3 masks. Int J Oral Maxillofac Surg 49:673–677. https://doi.org/10.1016/j.ijom.2020.03.015

Tasnim N, De la Vega L, Anil Kumar S, Abelseth L, Alonzo M, Amereh M, Joddar B, Willerth SM (2018) 3D Bioprinting Stem cell derived tissues. Cell Mol Bioeng 11:219–240. https://doi.org/10.1007/s12195-018-0530-2

Tedesco J, Lee BEJ, Lin AYW, Binkley DM, Delaney KH, Kwiecien JM, Grandfield K (2017) Osseointegration of a 3D printed stemmed titanium dental implant: a pilot study. Int J Dent 2017:5920714. https://doi.org/10.1155/2017/5920714

Tino R, Moore R, Antoline S, Ravi P, Wake N, Ionita CN, Morris JM, Decker SJ, Sheikh A, Rybicki FJ, Chepelev LL (2020) COVID-19 and the role of 3D printing in medicine. 3D Print Med 6:1–8. https://doi.org/10.1186/s41205-020-00064-7

Trenfield SJ, Awad A, Madla CM, Hatton GB, Firth J, Goyanes A, Gaisford S, Basit AW (2019) Shaping the future: recent advances of 3D printing in drug delivery and healthcare. Expert Opin Drug Deliv. https://doi.org/10.1080/17425247.2019.1660318

Turner RD, Wingham JR, Paterson TE, Shepherd J, Majewski C (2020) Use of silver-based additives for the development of antibacterial functionality in Laser Sintered polyamide 12 parts. Sci Rep 10:1–11. https://doi.org/10.1038/s41598-020-57686-4

van Hengel IAJ, Riool M, Fratila-Apachitei LE, Witte-Bouma J, Farrell E, Zadpoor AA, Zaat SAJ, Apachitei I (2017) Selective laser melting porous metallic implants with immobilized silver nanoparticles kill and prevent biofilm formation by methicillin-resistant Staphylococcus aureus. Biomaterials 140:1–15. https://doi.org/10.1016/j.biomaterials.2017.02.030

Verhoef LA, Budde BW, Chockalingam C, García Nodar B, van Wijk AJM (2018) The effect of additive manufacturing on global energy demand: an assessment using a bottom-up approach. Energy Policy 112:349–360. https://doi.org/10.1016/j.enpol.2017.10.034

VIT team aim to develop aptamer-based handheld biosensors for COVID-19 diagnosis—express healthcare (n.d.). https://www.expresshealthcare.in/news/vit-team-aim-to-develop-aptamer-based-handheld-biosensors-for-covid-19-diagnosis/424304/. Accessed 5 Oct 2020

Wang Y, Qi H (2020) Perfect combination of the expanded flap and 3D printing technology in reconstructing a child's craniofacial region. Head Face Med 16:1–5. https://doi.org/10.1186/s13005-020-00219-1

Wang H, Guo YT, Jiao Y, He DL, Wu B, Yuan LJ, Li YY, Yang Y, Cao TS, Zhang B (2019) A minimally invasive alternative for the treatment of nutcracker syndrome using individualized three-dimensional printed extravascular titanium stents. Chin Med J (Engl) 132:1454–1460. https://doi.org/10.1097/CM9.0000000000000255

Wesemann C, Pieralli S, Fretwurst T, Nold J, Nelson K, Schmelzeisen R, Hellwig E, Spies BC (2020) 3-D printed protective equipment during COVID-19 pandemic. Materials (Basel) 13:1–9. https://doi.org/10.3390/MA13081997

WHO Coronavirus Disease (COVID-19) Dashboard | WHO Coronavirus Disease (COVID-19) Dashboard (n.d.). https://covid19.who.int/. Accessed 11 Oct 2020

Willemsen K, Nizak R, Noordmans HJ, Castelein RM, Weinans H, Kruyt MC (2019) Challenges in the design and regulatory approval of 3D-printed surgical implants: a two-case series. Lancet Digit. Heal 1:e163–e171. https://doi.org/10.1016/S2589-7500(19)30067-6

Wong KC, Kumta SM, Geel NV, Demol J (2015) One-step reconstruction with a 3D-printed, biomechanically evaluated custom implant after complex pelvic tumor resection. Comput Aided Surg 20:14–23. https://doi.org/10.3109/10929088.2015.1076039

Xiu P, Jia Z, Lv J, Yin C, Cheng Y, Zhang K, Song C, Leng H, Zheng Y, Cai H, Liu Z (2016) Tailored surface treatment of 3D printed porous Ti6Al4V by microarc oxidation for enhanced osseointegration via optimized bone in-growth patterns and interlocked bone/implant interface. ACS Appl Mater Interfaces 8:17964–17975. https://doi.org/10.1021/acsami.6b05893

Xu N, Wei F, Liu X, Jiang L, Cai H, Li Z, Yu M, Wu F, Liu Z (2016) Reconstruction of the upper cervical spine using a personalized 3D-printed vertebral body in an adolescent with ewing sarcoma. Spine (Phila. Pa. 1976) 41(2016):E50–E54. https://doi.org/10.1097/BRS.0000000000001179

Yavari AS, van der Stok J, Chai YC, Wauthle R, Tahmasebi Birgani Z, Habibovic P, Mulier M, Schrooten J, Weinans H, Zadpoor AA (2014a) Bone regeneration performance of surface-treated porous titanium. Biomaterials. 35:6172–6181. https://doi.org/10.1016/j.biomaterials.2014.04.054

Yavari AS, Ahmadi SM, van der Stok J, Wauthle R, Riemslag AC, Janssen M, Schrooten J, Weinans H, Zadpoor AA (2014b) Effects of bio-functionalizing surface treatments on the mechanical behavior of open porous titanium biomaterials. J Mech Behav Biomed Mater 36:109–119. https://doi.org/10.1016/j.jmbbm.2014.04.010

Yavari AS, Croes M, Akhavan B, Jahanmard F, Eigenhuis CC, Dadbakhsh S, Vogely HC, Bilek MM, Fluit AC, Boel CHE, van der Wal BCH, Vermonden T, Weinans H, Zadpoor AA (2020) Layer by layer coating for bio-functionalization of additively manufactured meta-biomaterials. Addit Manuf 32:100991. https://doi.org/10.1016/j.addma.2019.100991

Zhang T, Zhou W, Jia Z, Wei Q, Fan D, Yan J, Yin C, Cheng Y, Cai H, Liu X, Zhou H, Yang X, Zheng Y, Liu Z (2018) Polydopamine-assisted functionalization of heparin and vancomycin onto microarc-oxidized 3D printed porous Ti6Al4V for improved hemocompatibility, osteogenic and anti-infection potencies. Sci China Mater 61:579–592. https://doi.org/10.1007/s40843-017-9208-x

Zhang Y, Geng X, Tan Y, Li Q, Xu C, Xu J, Hao L, Zeng Z, Luo X, Liu F, Wang H (2020) New understanding of the damage of SARS-CoV-2 infection outside the respiratory system. Biomed Pharmacother 127:110195. https://doi.org/10.1016/j.biopha.2020.110195

Zhao X, Xiao J, Sun Y, Zhu Z, Xu M, Wang X, Lin F, Wang Y, Wang J (2018) Novel 3D printed modular hemipelvic prosthesis for successful hemipelvic arthroplasty: a case study. J Bionic Eng 15:1067–1074. https://doi.org/10.1007/s42235-018-0094-9

Zhou P, Yang XL, Wang XG, Hu B, Zhang L, Zhang W, Si HR, Zhu Y, Li B, Huang CL, Chen HD, Chen J, Luo Y, Guo H, Di Jiang R, Liu MQ, Chen Y, Shen XR, Wang X, Zheng XS, Zhao K, Chen QJ, Deng F, Liu LL, Yan B, Zhan FX, Wang YY, Xiao GF, Shi ZL (2020) A pneumonia outbreak associated with a new coronavirus of probable bat origin. Nature 579:270–273. https://doi.org/10.1038/s41586-020-2012-7

Zhu Z, Dhokia V, Newman ST, Nassehi A (2014) Application of a hybrid process for high precision manufacture of difficult to machine prismatic parts. Int J Adv Manuf Technol 74:1115–1132. https://doi.org/10.1007/s00170-014-6053-7

Zhu FC, Li YH, Guan XH, Hou LH, Wang WJ, Li JX, Wu SP, Wang BS, Wang Z, Wang L, Jia SY, Jiang HD, Wang L, Jiang T, Hu Y, Gou JB, Xu SB, Xu JJ, Wang XW, Wang W, Chen W (2020) Safety, tolerability, and immunogenicity of a recombinant adenovirus type-5 vectored COVID-19 vaccine: a dose-escalation, open-label, non-randomised, first-in-human trial. Lancet 395:1845–1854. https://doi.org/10.1016/S0140-6736(20)31208-3

Fostering Novel Materials and Subsisting Technologies for 3D Printing

Himangi Sood, Simran Kaur, and Ishani Sharma

Abstract The COVID-19 pandemic made a drastic impact on the human health and overall economy of the world. Other than health-related issues, the global healthcare structure was deeply affected due to an alarming imbalance in the demand-supply chain of critical medical equipment such as ventilators, quarantine closets, oxygen valves, protective equipment, etc., that support patients' life during their battle against COVD-19. The field of 3D printing, or now commonly known as additive manufacturing (AM) has benefitted extremely by capitalizing on the increasing demand for manufacturing medical equipment. As a result, a vast number of start-ups have mushroomed up that are experimenting with various novel materials, techniques, and applications to build 3D printed products in quick turnaround time. Through this paper, we would highlight the propitious application of 3D printing in the biomedical engineering area. The chapter begins by addressing the fundamentals of 3D printing technologies and materials that shall be used to illustrate their whole system environmental impacts in later sections. This is followed by discussion over prominent countless initiatives that are being taken by multiple R&D centers, academic institutes, and various other corporations for the implementation of new technologies and materials to combat this pandemic. Further, the need for novel materials and the use of agricultural waste as compostable biomaterials for a circular economy has been discussed in detail. Aside to this, among the multiple 3D printing processes, a major focus has been on material extrusion, powder bed fusion, and material jetting techniques that are used for the medical sector based on current medical requirement, i.e., automatic door handle openers, face shields/masks, quarantine closets, ventilators, valves, etc. In the end, the conclusion will prove to be a stepping-stone for future research in this direction.

Keywords Biomedical · 3D printing · Three-dimensional printing · Additive manufacturing · 3D materials · Additive manufacturing · COVID

H. Sood (✉) · S. Kaur · I. Sharma
IDS Infotech Ltd, Mohali, Punjab 160062, India

K. Sandhu et al. (eds.), *Emerging Applications of 3D Printing During CoVID 19 Pandemic*, Lecture Notes in Bioengineering, https://doi.org/10.1007/978-981-33-6703-6_6

1 Introduction

3D printing is an additive manufacturing (AM) process of joining materials together to produce a product or part that takes the form of a 3D model. In today's time, 3D printing technology plays a major role in various industries—be it, aerospace engineering, biomedical engineering, automobiles, agriculture, food industry, or medical fraternity, etc. The year 2020 acted as an eye-opener when the COVID-19 pandemic encompassed the whole wide world, and we witnessed a steep increase in the demand for medical equipment—that too at a time when the entire supply chain had been drastically disrupted due to worldwide lockdowns imposed due to the onset of COVID-19 pandemic! This resulted in making the healthcare industry one of the major adopters of 3D printing technology. In the recent past, many new ventures came into existence that are constantly experimenting with various novel technologies, materials, and applications related to 3D printing or additive manufacturing within the healthcare industry. Such start-ups play a crucial role in driving the additive manufacturing industry—not only have they increased the rate at which biomedical equipment is manufactured, but also helped increase the sustainability and durability of the 3D printed products that are being used by the medical fraternity in our fight against the COVID pandemic. The novel 3D printing materials such as ceramic materials, biomaterials, electronic materials, smart materials, and composites are being considered to manufacture biomedical equipment nowadays. The properties of these novel materials should be analyzed in such a way that they meet the new requirements imposed on the products they are being used in, i.e., low energy printing, stable chemical/functional properties, etc. The novel 3D printing technologies aim at improving the existing state of the art technologies to minimize the requirement of the support structures, as well as to produce multiple layers in a single pass process to reduce the manufacturing time and increase the required stability. The two main factors that aid in efficient mass production of 3D printed goods by providing minimal risk for virus transmission through zero-contact process in quick turnaround time are:

- type of materials used
- type of 3D printing technology used.

The new start-ups are frequently experimenting with building new materials, such as smart materials, nanomaterial's, composite materials, or biomaterials, etc., that are capable of being used in 3D printing. However, with the onset of a pandemic, it has been observed that there is a need for advancement in 3D printing material type and technology type that can radically improve the whole system environment and help in further improving the mass production of bio-equipment that are used for providing protection and for testing of COVID-19.

2 Background

Many different types of AM methods are already known to mankind and are currently in use as well. These methods include, for instance, stereolithography (SLA), selective laser sintering (SLS), and digital light processing (DLP), fused deposition modeling (FDM), electronic beam melting (EBM), selective laser melting (SLM), droplet deposition/layering, and direct metal laser sintering (DMLS) (https://www.sculpteo.com/en/3d-printing/3d-printing-technologies/). These methods are selected and preferred keeping in mind the industry and type of devices that need to be manufactured. It is crucial to select the right set of materials and technology that must be used to produce the end output. The different technologies corresponding to different end-product requirement can be:

- SLA technique is used to develop the medical/dental products and electronics casings,
- SLS produces electronics housing mounts, custom consumer products, and aerospace hardware,
- The FDM technique builds the custom consumer products and electronics housing, aerospace ducting, automotive components, and food production tools, and
- DMLS is used in turbine engine components, instrumentation parts, medical equipment, and in automotive industries.

Table 1 illustrates the clear description of various conventional 3D printing technologies along with the widely used utilized materials.

The above mentioned techniques have various advantages; however, one advantage common to all the techniques is the formation of strong complex parts. FDM,

Table 1 Overview of 3D printing technologies

AM method	Technique implied	Preferred materials
Stereolithography (SRA)	UV light	Liquid, photopolymer
Fused deposition modeling (FDM)	Extrusion	Filament, polymers
Selective laser sintering (SLS)	UV light	Powder, polymers, metallic, ceramic
Electronic beam melting (EBM)	Electron beam	Powder, metallic
Direct metal laser sintering (DMLS)	Laser beam	Powder, metallic
Selective laser melting (SLM)	Laser	Powder, metallic
Digital light processing (DLP)	Light/energy Deposition	Photopolymers resin
Droplet deposition/layering	Inkjet	Powder along with liquid binder, polymer, ceramic, metallic

for instance, gives low cost and high strength parts whereas DLP is another low-cost technology with good accuracy with relatively quick turnaround time. SLA is another rapid fabrication technology that has the ability to create complex shapes with high resolution. However, with these advantages come certain disadvantages such as production of weak parts that are susceptible to sunlight and heat; grainy and rough surface finish which should also be considered while the selection of efficient technology and materials.

3 Novel Materials

Novel materials are the new advanced materials that have become critical components for new applications of various emerging technologies such as, but not limited to, 3D printing (Lee et al. 2017). We have clustered five types of novel materials further elaborated them in detail in detail below.

3.1 Composite Materials

Composite materials are the new materials in the field of 3D printing] (https://markforged.com/resources/learn/design-for-additive-manufacturing-plastics-composites/understanding-3d-printing-strength/3d-printing-carbon-fiber-and-other-composites; Blanco 2020; How composite materials are changing the world of additive manufacturing (again) 2017; Kalsoom et al. 2016). Composites typically comprise a core polymer material and a reinforcing material, like chopped or continuous fiber. The composite material offers higher strength and stiffness compared to non-reinforced polymers. In some cases, it can even replace metals like aluminum. The composite 3D printing has the ability to streamline and cut the cost of traditional composite manufacturing. There are numerous methods for fabricating composite components, in addition to 3D printing (Composite 3D Printing: an emerging technology with a bright future 2020). Few examples of these materials are present within AM process. Electronic sensors such as capacitance sensors and piezoresistive sensors are printed using the thermoplastic composite materials that enable the sensing of capacitance changes and mechanical flexing. These new composite materials can be used in additive manufacturing with FDM low-cost technology without any complex circuit requirement and high-cost resources of sensor production.

3.2 Biomaterials

Various types of implants are manufactured today that use biomaterials (Bandyopadhyay et al. 2015; Tappa and Jammalamadaka 2018; Jang et al. 2018). The implant fabricated using biomaterial is considered a success if it is biocompatible, inert, mechanically durable, and easily moldable which has made 3D printing technology revolutionary in medical and pharmaceutical fields. Mostly the biocompatible synthetic polymers, natural polymers, and acrylates-based polymers are preferred which are the recent developments in biocompatible materials that are being used to print the functional living tissues using AM process. These bio-printed tissues are used in organ transplantation that brings a revolution in the biomedical industry by addressing the challenge of organ transplantation.

3.3 Electronic Materials

Conventional AM technologies can fabricate the electronics components, such as resistors, capacitors, antenna, and inductors. The fabrication of flexible thin-film transistors has been reported on plastic substrates with self-synthesized ink and to achieve high resistance value using conducting polymer, resistors are printed on a plastic substrate. There are five materials that can be 3D printed with active properties and integrated into device components:

(1) An Elastomeric Matrix;
(2) Conducting Polymers;
(3) Quantum Dot Nanoparticles;
(4) A Uv-Adhesive Transparent Substrate Layer;
(5) Metal Leads;

Therefore, 3D printing is a flexible process to fabricate or manufacture electronic devices.

3.4 Ceramics Materials

This type of 3D printing uses materials such as clay or other fluid-dense materials for additive manufacturing processes. Compared with metal and polymers, ceramic materials have an extremely high melting point which is one of the biggest challenges in additive manufacturing. There is a new way to fabricate ceramics parts using 3D printing with particulars pre-ceramics monomers fusing with ultraviolet (UV) photo-initiator. Therefore, these 3D printed ceramics parts give excellent thermal stability without shrinkage (Chen et al. 2019; Ceramic 3D printer: all about ceramic 3D printing 2020; 3D printing with technical 2020). These ceramics materials are of great interest in porous burners, protection systems, electronic device packaging, propulsion components, and micro-electromechanical systems.

3.5 Smart Materials

Materials that have the capability to transform their shape or geometry under the impact of external parameters are known as smart materials (https://www.sculpteo.com/en/3d-learning-hub/best-articles-about-3d-printing/smart-materials/; Smart materials: 3D printing self-healing capsules for concrete 2019). 3D printing of these smart materials gives rise to 4D printing applications. 4D printing is a fledgling topic in the field of 3D printing where the fourth dimension is time. The basic principle of 4D printing is the use of programmable smart materials for 3D printing that can moderately change the shape over time based on external parameters, such as heat and water. One such example of this type of material is cellulose fibrils (nanocellulose).

All these various types of new materials explained above play a major role in 3D printing's long term future. Table 2 represents a summary of the novel materials and their utilization in the 3D printing field.

Table 2 Novel materials in 3D printing (Lee et al. 2017)

Novel materials category	Materials	AM process	Application
Smart materials	Shape memory polymers	Vat photopolymerization	Actuator, sensor, jewelry, gripper
Ceramic materials	UV curable monomers	Vat photopolymerization	Thermal protection
Electronic materials	Silver nanoparticle ink, conductive polymer, quantum dot	Material jetting	Thin-film transistor, antenna emitter, resistors, LEDs
Composite materials	CB/PCL	Material extrusion	Sensors
	Verowhite Plus and Tangoblack Plus	Material jetting	Fracture resistant composites
	Barium Titanate Nanoparticle/Polyethylene Glycol Diacrylate	Vat photopolymerization	3D piezoelectric polymers
Biomaterials	Hydrogels, functional inks	Material extrusion	Tissue engineering, cardiac micro-physiological devices

4 Novel Technologies

In today's time, the most widely used technique for the production of 3D printing devices is Fused Deposition Modeling (FDM). It is a type of low-cost material extrusion process. A wide range of materials can be used in this technique. However, there is a need for new 3D printing technologies as FDM suffers from slow speed and sometimes also leads to the problem of warping.

Recently, new 3D printing techniques are being researched upon that would eradicate the existing disadvantages of conventional 3D printing techniques. These are listed herein and further explained in detail below: (1) computed axial lithography (CAL) (Davide 2019) (Tetsuka and Shin 2020), (2) continuous liquid interface production (CLIP) (Service RF 2019), (3) rotational 3D printing (Novel 3D printing technique yields high-performance composites Arranging fibers just like nature does 2018), (4) inverted multi-material laser sintering (Hanaphy 2020).

4.1 Computed Axial Lithography (CAL)

The 3D printing technique that is inspired by computed tomography (CT) is called computed axial lithography (CAL). This technique is based on photopolymerization and utilizes photo-responsive material placed in a liquid made from polymer and is subjected to CT scans to obtain a flexible and complex 3D object which later is crafted to the desired shape by projecting light to a rotating cylinder. Through the utilization of CAL, the support structure that is a major requirement of conventions 3D printing techniques can be circumvented. This technique also eliminates the requirement for supports and one can build complex and nested structures that were previously challenging or nearly impossible to print. The process is almost layer-less without involving any relative motion between resin and printed part—thereby providing immense strength to the structures produced via this technique (Bhattacharya et al. 2018).

4.2 Continuous Liquid Interface Production (CLIP)

The 3D printing technique is 100X faster! This technique is also based on photopolymerization, where ultraviolet light is projected inside a space filled with liquid resin which is circulated with a liquid coolant from beneath to reduce the effect of warping and cracking of the final 3D printing product (Janusziewicz et al. 2016; Johnson et al. 2016). A continuous liquid interface production (CLIP) is achieved with an oxygen-permeable window below the ultraviolet image projection plane The CLIP creates a "dead zone" (persistent liquid interface) where photopolymerization is inhibited between the window and the polymerizing part. We delineate critical control parameters and show that complex solid parts can be drawn out of the resin at rates of hundreds of millimeters per hour. These print speeds allow parts to be produced in minutes instead of hours (Tumbleston et al. 2015).

4.3 Rotational 3D Printing

Rotational 3D printing is a new technology that aids in making stronger parts! This additive manufacturing programs the fiber orientation within epoxy composites in specified locations to generate structures with characteristics of high strength and stiffness (A new spin on 3D printing can produce an object in seconds 2020; Rotational 3D printing: a new technology for stronger parts 2018; Raney et al. 2018). This is done with the help of a spinning nozzle—hence the name rotational printing. Since the process allows us to modify the complex microstructures of the material, therefore, the parts produced have higher strength and stiffness, thereby increasing damage tolerance.

4.4 Inverted Multi-material Laser Sintering

Selective laser sintering (SLS) is a traditional process for 3D printing a metal that uses a laser beam to focus upon a print bed filled with metal powder. When the laser beam moves back and forth in a pre-defined pattern, it melts and fuses the powder into a solid at selected areas. The product is built up gradually, one layer at a time, with the addition of more powder. However, the switching of two metals back and forth is a difficult process, especially when both the metals have different melting points (Upside-down 3D printing tech creates multi-material 2020).

Inverted multi-material laser sintering technique is a new and improved version of SLS that reduces the steps for printing by sintering multiple powders materials of thermoplastic and metal in a single run. This is achieved through inverting the laser inside the 3D printer and replacing the powder bed box with multiple glass plates. Through this new technique, embedded circuit boards and robotic components could be easily printed.

5 Conclusion and Future Scope

We believe the world is forever changing and medicine is always going to be one area where research is bound to always continue. The scope of additive manufacturing in the medical industry is limitless. With the advancement in medical technology, people have been living longer. The medical fraternity is benefiting hugely through the use of 3D printing, where the outcome is truly astonishing and remarkable. We can now think of creating 3D printed organs and other medical equipment in a quick turnaround time. To conclude, we believe that this pandemic has really helped us identify 3D printing as the most viable option that comes with many advantages as it can provide quicker and better alternative solutions to most of the traditional processes. With the ability to apply 3D printing into almost any industry very amicably, the evolution of this technology will always remain an ongoing task as we constantly strive for developing better materials and new and improvised methods to yield unimaginable outcomes in the shortest time-span. With the advent of all these novel materials and technologies in 3D printing, every industry can take benefits in their object manufacturing process, especially the biomedical industry as 3D printing technology can address the major issues they are facing during this COVID-19 pandemic time. For example, in Italian hospitals, as the number of Coronavirus patients increased, they required breathing machines and on the other side, the original suppliers were unable to fulfill this sudden high demand for oxygen valves. So, in response to this situation, an engineering firm of Brescia started implementing 3D printing to meet the requirements of hospitals and this step has saved many patients lives (Petch 2020).

In the current scenario of COVID, a major need of every person may be a healthcare worker or patient or any other person who wants to stay protected from the virus is personal protective equipment (PPE). Out of the long list of PPE's such as face shield, masks, respirators gloves, goggles or glasses, gowns and head cover, masks or respirators remains a top priority requirement for all. As COVID-19 is carried mainly through the association with an infected person, the major issue faced while using these products is virus transmission when an infected person touches the final end-product, and the virus stays over the surface for a certain time. Table 3 highlights

Table 3 Lifetime on different surfaces

Type of surface	Virus stay time
Plastic	3–4 days
Polypropylene plastic	3 days
Metal	5 days
Paper	3 h
Cardboard	24 h
Ceramic	5 days
Aluminum	2–8 h
Stainless steel	2–3 days
Copper	4 h
Glass	4 days
Outside of surgical mask	7 days
Surgical glove	8 h

different stay time of the COVID virus on different surfaces used (https://www.bigstockphoto.com/image-368913025/stock-vector-coronavirus-infographic-lifetime-of-covid-19-virus-infection-on-different-surfaces-and-materials%2C-d; https://www.webmd.com/lung/how-long-covid-19-lives-on-surfaces; https://scroll.in/article/957211/dirty-money-can-coronavirus-spread-through-bank-notes). As can be referred from Table 3, for the basic surgical mask, the lifetime stay of the virus is 7 days; therefore, these cannot be used again and again and hence require optimization in the supply of these. Also, these surgical masks do not provide complete protection from bacteria or viruses due to inherent loose fit. N95 masks, which are the reusable form of masks with tight-fitting, use steel for staples and aluminum for nose clip, which also inherent virus lifetime for a minimum of 2–3 days. These issues can be resolved by using 3D printed products, but for that certain improvements are also required in the materials and the technologies.

For the optimization of supply, 3D printed face masks are in production, however, the major guidelines issued by FDA for surgical masks and respirators such as filtration and infection control are not effectively provided by these 3D printed face masks. Many individuals are working in this direction so as to maximize filtration support and reusability of these 3D printing masks by including special additive materials during 3D printing such as biocompatible polymer containing a copper nanocomposite (https://copper3d.com/hackthepandemic/) or silver ions (https://3dprint.com/273171/3dfils-introduces-line-of-silver-ion-antibacterial-3d-printing-filaments/) or NaCl filter as a middle filter (Sachan 2020). Although the role of these materials is not yet known for COVID, yet these materials provide additional antimicrobial properties to the mask which should also be well investigated for efficient results (Table 4).

Also, to solve the loose fit issue for masks, research is being focused on customizing mask seal design (Mang et al. 2018). To fix this 3D laser scanning for scanning facial parameters such as face and nose length and structure to generate N95 template can be utilized which can further be used with Fused Deposition Modeling 3D printer utilizing Acrylonitrile Butadiene Styrene plastic.

Apart from the PPE kits, the other area where most of the research is being focused on is 3D printed quarantine booths, which have also witnessed a shortfall due to the rise in cases of COVID around the world. The solid urban construction where waste material as a raw material is used could provide rigid support to the structure of quarantine booth (Petch 2020), however, the limitation faced here is higher expenses faced and size issue. Therefore, it is impertinent to investigate more on the type of materials that can be further utilized in this scope to eradicate the abovesaid disadvantages. Aside to this, another structure, i.e., 3D printed testing booths in parallel to the utilization of Fiberglass composites as raw material for building a lighter and sturdier structure can be a hot topic of future research.

As mentioned previously, FDM is a priority choice for producing 3D printing products and COVID products are no exception here (Mang et al. 2018).

Apart from the FDM technique, research should also be aligned towards the possibility of finding new techniques which utilize object formation inside a liquid

Table 4 3D printed examples, reported in the news, in the fight against COVID-19 (Mang et al. 2018)

AM technology	AM product for supporting healthcare professional
FDM	Hand sanitizer holder
FDM	Door handle attachment
FDM	Screw less hand-free door handle openers
FDM	Vision/face shield
FDM	Face masks (E.G. surgical & N95 respirator)
FDM	Non-invasive positive end expiratory pressure (PEEP) masks
FDM	3D printed quarantine booths
FDM	Ventilators
FDM	Parts to convert existing manual ventilator systems into automatic ones
FDM	Adaptors for a variety of medical devices
SLS & FDM	Oxygen valves
SLA	Venturi type valves for respirators (Needs Regulatory Approval)
SLA	COVID-19 test swabs
SLA	3DP lung models for use in surgical planning & understand COVID-19

which would result in a germ-free end-product—a high priority requirement during the COVID pandemic.

One of the other techniques is computed axial lithography (CAL). The computed axial lithography can build a full 3D structure in a single step using liquid resin as it works on sequential illumination from many different angles using series of 2D images. It has an added advantage that it can build the objects very quickly—a quality that is much sought after, given the current situation. And there is another technique, continuous liquid interface production (CLIP) that utilizes liquid resins beneath which liquid coolant is placed and can print a huge structure in a limited time.

References

3D printing with technical ceramics (2020). https://www.beamler.com/3d-printing-with-technical-ceramics/

A new spin on 3D printing can produce an object in seconds (2020). https://arstechnica.com/gadgets/2020/02/a-new-spin-on-3d-printing-can-produce-an-object-in-seconds/

Bandyopadhyay A, Bose S, Das S (2015) 3D printing of biomaterials. MRS Bull 40(2):108–115

Bhattacharya I, Kelly B, Shusteff M, Spadaccini C, Taylor H (2018) Computed axial lithography: volumetric 3D printing of arbitrary geometries (Conference Presentation). In: Image sensing technologies: materials, devices, systems, and applications, vol 10656. International Society for Optics and Photonics, p 106560P

Blanco I (2020) The use of composite materials in 3D printing. J Compos Sci 4(2):42

Ceramic 3D printer: all about ceramic 3D printing in 2020. https://all3dp.com/2/ceramic-3d-printer-ceramic-3d-printing/
Chen Z, Li Z, Li J, Liu C, Lao C, Fu Y, Liu C, Li Y, Wang P, He Y (2019) 3D printing of ceramics: a review. J Eur Ceram Soc 39(4):661–687
Composite 3D Printing: an emerging technology with a bright future. https://amfg.ai/2020/02/25/composite-3d-printing-an-emerging-technology-with-a-bright-future/
Coronavirus infographic. Lifetime of COVID-19 virus infection on different surfaces and materials, dangerous pneumonia precaution. Vector info virus viability. https://www.bigstockphoto.com/image-368913025/stock-vector-coronavirus-infographic-lifetime-of-covid-19-virus-infection-on-different-surfaces-and-materials%2C-d
Davide S (2019) More details emerge on UC Berkeley-LLNL new CAL volumetric 3D printing. https://www.3dprintingmedia.network/more-details-emerge-on-uc-berkeley-llnl-new-cal-volumetric-3d-printing-method/
Dirty money: can coronavirus spread through bank notes?. https://scroll.in/article/957211/dirty-money-can-coronavirus-spread-through-bank-notes
Hack the pandemic. https://copper3d.com/hackthepandemic/
Hanaphy P (2020) Columbia researchers develop a new method for multi-material Sls 3D printing. 3D Print Ind. https://3dprintingindustry.com/news/columbia-researchers-develop-a-new-method-for-multi-material-sls-3d-printing-174108/
How composite materials are changing the world of additive manufacturing (again) (2017). https://www.3dprintingmedia.network/composite-materials-changing-world-additive-manufacturing/
How long does the coronavirus live on surfaces?. https://www.webmd.com/lung/how-long-covid-19-lives-on-surfaces
Jang J, Park JY, Gao G, Cho DW (2018) Biomaterials-based 3D cell printing for next-generation therapeutics and diagnostics. Biomaterials 1(156):88–106
Janusziewicz R, Tumbleston JR, Quintanilla AL, Mecham SJ, DeSimone JM (2016) Layerless fabrication with continuous liquid interface production. Proc Natl Acad Sci 113(42):11703–11708
Johnson AR, Caudill CL, Tumbleston JR, Bloomquist CJ, Moga KA, Ermoshkin A, Shirvanyants D, Mecham SJ, Luft JC, DeSimone JM (2016) Single-step fabrication of computationally designed microneedles by continuous liquid interface production. PLoS ONE 11(9):e0162518
Kalsoom U, Nesterenko PN, Paull B (2016) Recent developments in 3D printable composite materials. RSC Adv 6(65):60355–60371
Lee J-Y, An J, Chua CK (2017) Fundamentals and applications of 3D printing for novel materials. Appl Mater Today 7:120–133
Mang C, Li H, Shen S, Wang Y, Yang Q (2018) Customized design and 3D printing of face seal for an N95 filtering facepiece respirator. J Occup Environ Hyg 15(3):226–234
Novel 3D printing technique yields high-performance composites Arranging fibers just like nature does it (2018). https://www.seas.harvard.edu/news/2018/01/novel-3d-printing-technique-yields-high-performance-composites
Petch M (2020) 3D printing community responds to COVID-19 and coronavirus resource. https://3dprintingindustry.com/news/3d-printing-community-responds-to-covid-19-and-coronavirus-resources-169143/
Raney JR, Compton BG, Mueller J, Ober TJ, Shea K, Lewis JA (2018) Rotational 3D printing of damage-tolerant composites with programmable mechanics. Proc Natl Acad Sci 115(6):1198–1203
Rotational 3D printing: a new technology for stronger parts (2018). https://www.sculpteo.com/blog/2018/09/21/rotational-3d-printing-a-new-technology-for-stronger-parts/
Sachan D (2020) COVID-19 pandemic has spurred materials researchers to develop antiviral masks 98(31)
Service RF (2019) Fast new 3D printing method creates objects as big as an adult human. https://www.sciencemag.org/news/2019/10/fast-new-3d-printing-method-creates-objects-big-adult-human

Smart materials: 3D printing self-healing capsules for concrete (2019). https://3dprint.com/248636/smart-materials-3d-printing-self-healing-capsules-concrete/

Smart materials: what is it and how to use them?. https://www.sculpteo.com/en/3d-learning-hub/best-articles-about-3d-printing/smart-materials/

3D Printers and 3D Printing: Technologies, Processes and Techniques. Sculpteo. https://www.sculpteo.com/en/3d-printing/3d-printing-technologies/

3D printing carbon fiber and other composites. https://markforged.com/resources/learn/design-for-additive-manufacturing-plastics-composites/understanding-3d-printing-strength/3d-printing-carbon-fiber-and-other-composites

Tampi T, 3DFils introduces line of silver ion antibacterial 3D printing filaments. https://3dprint.com/273171/3dfils-introduces-line-of-silver-ion-antibacterial-3d-printing-filaments/

Tappa K, Jammalamadaka U (2018) Novel biomaterials used in medical 3D printing techniques. J Funct Biomater 9(1):17

Tetsuka H, Shin SR (2020) Materials and technical innovations in 3D printing in biomedical applications. J Mater Chem B 8(15):2930–2950

Tumbleston JR, Shirvanyants D, Ermoshkin N, Janusziewicz R, Johnson AR, Kelly D, Chen K, Pinschmidt R, Rolland JP, Ermoshkin A, Samulski ET (2015) Continuous liquid interface production of 3D objects. Science 347(6228):1349–1352

Upside-down 3D printing tech creates multi-material objects (2020). https://newatlas.com/3d-printing/upside-down-3d-printing-multi-material-objects/

COVID-19 Safe Clinical Dentistry: Applications of 3D Printing Technologies

Anoop Kapoor, Ishdeep Kaur, Arshdeep Kaur, Himanshu Deswal, Rajni Jain, Prabhjot Kaur, and Vishakha Grover

Abstract Dental profession is at a very critical conjecture owing to novel coronavirus sickness, COVID-19. Dentistry is a profession where healthcare professionals are extremely susceptible of being infected as well as retain a high susceptibility to cross infect and transmit the contagious viral sickness among general public attending dental clinics. The grounded reality for an average dental practitioner is that walking back to operative clinical work in dentistry post-pandemic is quite tedious. Most of the safeguards and practices recommended in the current COVID-19 pandemic phase need significant alterations in the ways to practice clinical dentistry in terms of prevention and transmission of cross infection. 3D printing or additive manufacturing is the most upcoming surge of technology in the field of health care. Many novel research-driven and technology-based interventions from 3D printing technologies must be exploited and adapted to find ways and means to fight out the outburst of the pandemic. Many such 3D printed mass produced and cost-effective tools, equipments, and barrier materials instilled in clinical dental practice shall prevent or minimize COVID-19 cross contamination and transmission by ensuring safe clinical practices.

Keywords 3D printing · COVID-19 · Pandemic · Cross infection · Personal protective equipment · Antibacterial

---

A. Kapoor
Sri Sukhmani Dental College and Hospital, Dera Bassi, Punjab, India

I. Kaur · A. Kaur · R. Jain · P. Kaur · V. Grover (✉)
Dr. H.S.J. Institute of dental sciences, Panjab University Chandigarh, Chandigarh, India

H. Deswal
Dental Officer, ECHS-Polyclinic, Ropar, Punjab, India

K. Sandhu et al. (eds.), *Emerging Applications of 3D Printing During CoVID 19 Pandemic*, Lecture Notes in Bioengineering, https://doi.org/10.1007/978-981-33-6703-6_7

1 Introduction

It has been very recent since the world had first heard the word "Sars-CoV2 virus", and the novel coronavirus has changed the phrase "talk of the town" to the "talk of the globe". It is a malevolent silent intruder and has impacted more than 213 nations worldwide (COVID-19 coronavirus pandemic 2020).

The world undoubtedly is battling its worst health emergency in the form of the COVID-19 pandemic, and the scale of situation is such that many nations have extended the nationwide lockdowns as the key strategy to stop the progression in the absence of any known reliable remedy for coronavirus infection. The lockdowns have served the purpose to some extent, as bought time to be prepared to handle the emergent situation, managing the cases affected by fatal viral infection, upgrading our health infrastructure, flattening the curve of transmission, speeding up testing strategy, and a much fast-paced research for diagnosis and management of novel coronavirus infection. The percussions of the pandemic shall sound tough on each and every sector, but there will be some arenas which shall bear the hardest brunt.

Dental profession at this time point is at a very critical conjecture owing to novel coronavirus sickness, COVID-19. Dentistry is a profession which is extremely susceptible of being infected as well as with highest susceptibility to cross infects and transmit the contagious viral sickness among general public attending dental clinics. Even with loads of guidelines, recommendations, regulations for combating COVID-19 available from various reputed and responsible healthcare organizations in the country and worldwide, the grounded reality for an average dental practitioner in developing countries is that walking back to clinical work is still tedious. Most of the safeguards and practices recommended in this emergence phase of pandemic need significant alterations in the ways to practice clinical dentistry. The standard sterilization and disinfecting protocols, clinical operative work procedures, dental instruments, and equipments along with preparation of dental clinics for post-COVID-19 dentistry is going to take out a big chunk of investment from the practicing dentists. Majority of dentists practicing in rural and suburban areas, running very minimalistic equipped dental clinics, merely to cater to very basic dental care needs of the community around may not be able to bear the financial burden of upgrading the operatory and practices to make them corona proof and continue to serve. Even dental practitioners who may cope up, the availability of these recommended materials, instruments, and equipments past lockdown phase such as non-contact thermal sensors, improvised barrier materials (personal protective equipments and N95 respirators) and effective viricidal disinfectants and antiseptics, high vacuum suction apparatus, etc., is not easy, owing to limited import and indigenous production. Next, ensuring the affordability and availability, adaptability of the Indian dentist population is another barrier. It shall definitely be a varied learning curve to utilize and tailor these new tools and techniques into existing clinical practices by dentists to safeguard themselves including their families and patients attending their clinics.

Unlike many other healthcare streams, dentists are to work in close contact with their patients, for a protracted period of time, as almost all dental treatments are operative procedures. Barely a dental patient can be fully treated solely by the prescription of a medicine. So, there is a very high chance of coming across asymptomatic patients, who may be highly susceptible of transmitting infection or even COVID positive status. What makes the things worse is that the coronavirus is new, yet unknown, undergone mutational changes, and most uniquely a highly transmissible agent, having a high affinity to oral tissues and saliva. A cell surface receptor on oral epithelial cells of humans viz angiotensin-converting enzyme 2 (ACE2) is vital for COVID-19 infectivity as it binds to the viral attachment protein, the spike (S) glycoprotein, critical for diseases transmission (Xu et al. 2020a, b). Approximately 91.7% positive rate of occurrence of COVID-19 including live virus in patients' saliva has been observed (To et al. 2020). COVID-19 can be transmitted from infected saliva even in case of asymptomatic cases, which is an alarming concern for dental practitioners. Thus, in the truest sense, it is very strongly recommended to practice only safe clinical dentistry, with adequate preparation to combat novel coronavirus transmission. This is imperative as dental clinics start esp. in the post-lockdown period and until we are equipped with an effective and validated therapeutic agent against coronavirus.

2 COVID-19 and Post-pandemic Work Practice Recommendations for Dental Workers (Adapted from New Guidelines by OSHA and CDC)

2.1 General Aspects

Coronaviridae family to which coronaviruses belong harbors a single-stranded RNA as the viral genome (Fehr and Perlman 2015; Fan et al. 2019). Four major types of coronavirus have been known—alpha, beta, gamma, and delta (Perlman and Netland 2009). Humans are primarily affected by alpha- and beta-coronaviruses which target their respiratory, central nervous, and gastrointestinal functions (Fehr and Perlman 2015; Weiss and Leibowitz 2011; Yin and Wunderink 2018; Wu et al. 2020). Human angiotensin-converting enzyme 2 (ACE-2) receptor on which both 2019-nCoV and SARS-CoV can attach. ACE-2 receptor of humans, civet cat, pig, and bat on which 2019-nCoV can attach but in absence of ACE-2 receptor it is unable to attach to any other cells (Zhou et al. 2020; Peng et al. 2020; COVID-19symptoms may include altered senses of smell, taste [internet] 2020). With strong co-relation between 2019-nCoV and ACE-2 receptor, S protein indicated that individual with greater expression for ACE-2 is more susceptible to nCoV-2019 (COVID-19symptoms may include altered senses of smell, taste [internet] 2020).

2.2 Clinical Presentations

COVID-19 patient shows the following symptoms like cough, fever, myalgia (muscle pain), and shortness of breath, altered computed tomographic patterns human chest, tiredness and less severe ones like production of sputum, headache, gastric pain, hemoptysis, dizziness, vomiting, and diarrhea, etc. Some doctors might consider symptoms for COVID-19 like alteration in smell blindness and taste sensations. Onset of COVID-19 causes impairment of alveolar tissues in lungs and may cause respiratory and even death (Xu et al. 2020; COVID-19 symptoms may include altered senses of smell, taste [internet] 2020; del and Malani 2020; Huang et al. 2020).

2.3 COVID-19 Transmission

Epidemiological and genetic studies reveal that COVID-19 can spread from animal to humans initially and later on spread from humans to humans (del and Malani 2020; Lu et al. 2020). Classic pathway for spreading of COVID-19 is cough, inhalation of aerosol, sneeze and direct contact with oral cavity, nasal mucous membrane, and eye. Coronavirus transmission is not limited to respiratory tract only, rather open body cavities, e.g., eye, oral cavity, etc., may also serve as another portal of viral transmission. Coronavirus is likely to be spread via saliva indirectly or directly (Huang et al. 2020; Lu et al. 2020; To et al. 2020; Belser et al. 2013). Symptomatic patients who are infected by coronavirus are the primary source of infection, and those patients which are asymptomatic or those in the incubation period are carriers. Moreover, there is yet no clear evidence as to how the infected patients may be transmitting the infection during recovery and to what extent (Rothe et al. 2020). Many studies indicate that coronavirus may be airborne due to the generation of aerosol during medical procedures (Wax and Christian 2020). Aerosol spreading is one of the potent transmitting pathways in closed areas. During dental procedures, generation of aerosol is one of the potent sources of infection among dental staff and patients. Droplets origin may be either oropharyngeal or nasopharyngeal, which is usually linked with saliva. Virus spreads through large droplets affecting the smaller area, i.e., nearby area, while small droplets affecting larger area, i.e., long-distance area by suspended virus particles in the air (Xie et al. 2009). As consequence, the dental team should be very observant toward health of patients and themselves.

2.4 Incubation Period

Incubation period for coronavirus has been estimated to be 5–6 days, which can last up to 14–17 days, which is a broadly accepted range for quarantine and medical examination of highly infected individuals (Meng et al. 2020).

2.5 Diagnosis

Diagnosis of coronavirus relies on the travel history either to infected area or resides in infected area 2 weeks prior the initiation of the symptoms, CT scan results, laboratory tests, clinical symptoms. If the outcome of test is single negative that doesn't mean the suspected individual is not infected with virus. We should always co-relate history of individual, positive CT scan results, and COVID-19 symptoms (Meng et al. 2020). It has been noticed that coronavirus is present in the saliva of the infected individuals, therefore, health professionals dealing with biologic fluids are at high risk and they should be careful regarding virus transmission. Diagnosis of coronavirus may be carried out by monitoring viral counts in saliva. Some strains of virus stay in saliva for long periods after having infection (29 days) (Barzon et al. 2016; Zuanazzi et al. 2017), proposing that it could be novel method to analyze differences in disease-associated biomarkers in a non-invasive manner by salivary diagnostics (Segal and Wong 2008).

NCoV-2019 presence in the saliva can be proved by 3 discrete pathways which are as follows (Sabino-Silva et al. 2020):

1. nCoV-2019 from the respiratory tract have a communication into the oral cavity along with the droplets exchange associated during breathing.
2. nCoV-2019 present in blood stream enters into mouth via crevicular fluid.
3. Salivary gland infection with particles release subsequently into saliva via salivary ducts.

Epithelial cells of the salivary gland are affected with SARS-CoV in a short span as observed in case of rhesus macaque, so it is speculated that salivary gland cells may be acting as a prime source for transmission of virus via saliva (Sabino-Silva et al. 2020). Additionally, it has been revealed earlier that specific secretory antibodies (SIgA) against growth of SARS-CoV have been observed in saliva in animal models' studies (Lu et al. 2010). Due to the existence of diverse viral strains, salivary diagnosis of 2019-nCoV using specific antibodies against the virus may be developed as a differentiating strategy. Saliva can play a key role in human transmission of virus, but at the same time, it may be utilized as a tool for non-invasive, cost effective, and point of care diagnosis for infection with virus also (Sabino-Silva et al. 2020).

2.6 Prevention of Infection Transmission in Dental Clinic

We can manage COVID-19 by contacting patients earlier to dental procedure by using telephone screen asking for the symptoms and if the patient has symptoms of coronavirus, then refrain non-emergency dental treatment and if possible, hamper dental treatment until the recovery of patient and if possible we use teledentistry option for consultation. Prior appointments should be taken in case of emergency dental treatment and also advice the patient and accompanying person, the mandatory wearing of a face mask when they enter into the premises and assessed for the COVID-19 symptoms and fever, etc. (https://www.cdc.gov/coronavirus/2019-ncov/hcp/dental-settings.html; https://www.osha.gov/Publications/OSHA4019.pdf).

2.6.1 Work Practices

If possible clinical care should be limited to one person at given point of time by dental healthcare personnel. Sterile or clean instruments and supplies whichever is required for the dental treatment are easily approachable in the clinical settings. Instruments should be stored in covered places like cabinets or drawers so that they can be prevented from contamination. And all those instruments and supplies which are exposed to the environment and not used during treatment may be presumed that they are contaminated and should be dumped or sterilize properly for next use after finishing of procedure. Try to refrain from generation of aerosol during treatment. And also refrain from the use of 3-way syringe and handpiece. Ultrasonic scalers usage is also not suggested. Preference will be given to ATR (atraumatic restoration) techniques so that less aerosol generation is there. If it is necessary to perform aerosol-generating procedures in the clinic, then ways to minimize the production and best management protocols for aerosol and droplet splash are four-handed dentistry, high evacuating suction, use of rubber dams, etc. Number of dental healthcare personnel should be limited to required once so that less exposure to others (https://www.cdc.gov/coronavirus/2019-ncov/hcp/dental-settings.html; https://www.osha.gov/Publications/OSHA4019.pdf).

2.6.2 Hand Hygiene

Dental healthcare personnel should strictly adhere to proper hand hygiene when they are in contact with patients before and after procedure, contact with highly infectious materials, before don and after doff of PPE kit, including mask and gloves. After doffing of PPE kit, hand hygiene is of utmost importance for removal of any pathogen which might be associated with them. Use alcohol-based hand rub with 60–95% alcohol or wash hands for not less than 20 s with soap and water. For cleaning soiled hands, a wash with detergent and water is necessary, before the use of alcohol-based hand rubs or decontamination (https://www.cdc.gov/coronavirus/2019-ncov/hcp/dental-settings.html; https://www.osha.gov/Publications/OSHA4019.pdf).

2.6.3 Using Personal Protective Equipment (PPE)

PPE kit given to dental healthcare personnel should be in consonance with the Occupational Safety and Health Administration by employers. Dental healthcare personnel must be trained about when to use, how to don and doff personal protective equipment kit in order to restrain themselves from contamination, how to disinfect or dispose of and maintain, and also about limitations of personal protective equipment kit. Reusable PPE kit should be decontaminated cleaned after use and should be maintained properly after and between uses. For safe donning and doffing of PPE kit, dental setting should have certain policies and procedure which describes the sequence. Dental healthcare personnel should wear surgical mask, eye protective equipment, gown during treatment where chances of splattering of blood or body fluids are there. While doing procedures on patients in which aerosol is generating assumed to be non-contagious, usage of respirators or N95 respirators that provides extreme level of protection such as elastomeric respirators, PAPRs, or other disposable filtering facepiece respirators, if available. Respirators should be used well in adherence of respiratory protection protocol, which involves fit testing, training, and medical evaluations. If respirators are unavailable while doing aerosol-generating procedures use full face shield and surgical mask (US Food and Drug Administration (FDA) approved) and if full face shield and surgical mask are unavailable then avoid aerosol-generating procedures (https://www.cdc.gov/coronavirus/2019-ncov/hcp/dental-settings.html; https://www.osha.gov/Publications/OSHA4019.pdf).

2.6.4 Environmental Infection Control

Dental healthcare personnel must disinfect and clean the operatory after performing patients or in between patients so that environment should be cleaned and less chances of spread of infection although it is not mandatory to sterilize operatory between patients. For cleaning and disinfection of the operatory after a patient without confirmed or suspected coronavirus, wait for at least 15 min after termination of procedures and exit of patient from clinic to initiate to disinfect or clean room surfaces. During this time period droplets which are suspended in the air fall down sufficiently and then area to be disinfected satisfactorily. Regularly disinfecting and cleaning firstly with cleaner and water for cleaning of the surface and then use of Environmental Protection Agency-registered, hospital-grade disinfectant shall be practiced for commonly touched objects or surfaces. The disinfection protocol should be adhered to suitable exposure time and with products active against SARS-CoV-2 in healthcare settings. Standard and quality approved disinfectants that have been recommended by national regulatory agencies for use against SARS-CoV-2 should be utilized for best practices (https://www.cdc.gov/coronavirus/2019-ncov/hcp/dental-settings.html; https://www.osha.gov/Publications/OSHA4019.pdf).

3 3D Printing—As Potential Novel Solution During COVID-19 Pandemic

An adaptable, robotic platform permitting for customized deposition of biomaterials using computer-aided design (CAD) systems to compose layer-by-layer custom designs with controlled composition and architecture is called three-dimensional printing (Ishack and Lipner 2020; Cai et al. 2018; Banerjee et al. 2019; Bachtiar et al. 2020; Yu and Chen 2020). There is shortage in supply of face shield, N95 mask, test kits, PPE kit, and ventilators valves during this pandemic period (Ranney et al. 2020; Marco et al. 2020). Sufficient production and supply of PPE kit are of utmost importance for safe practice in dentistry during this pandemic period. An innovative and novel technique, i.e., 3D printing is used for fabrication of complex architecture to compete for the shortage.

Conventionally, subtractive technique is used by the industries in which solid block was carved out for the construction of products. In contrast, a layer-by-layer technique is used for additive fabricating that permits for interior modeling and complex designs. First uses of additive manufacturing were rapid prototyping which permits the producer to produce prototypes swiftly, assisting in assessing and testing of designs before final production of the product. In rapid prototyping, first three-dimensional models are fabricated by using computer-aided design software and then machine build three-dimensional objects based on that model (Amanda and Al'Aref 2018).

3.1 Evolution of 3D Printing as an Emerging Mass Production Technique

Evolution of 3D printing way back to 1960 at Battelle Memorial Institute in Ohio where research is conducted by using photopolymers for the creation of 3D objects. The aim of this experiment was to polymerize the resin by using two intersecting laser beams of different wavelengths. Solid photography was invented by Dynell Electronics Corporation in the late 1970s. In this technology, cross sections are cut based on computer model by using milling and laser machine, and then pile them together to form an Object (Wohlers and Gornet 2014). Wyn Swainson applied a patent for dual laser-based methods in 1971, which were quite similar to photochemical machining (https://www.google.com/patents/US4041476). He further discovered the Form graphic Engine Company in California, but the technology failed to be commercialized successfully (Beaman et al. 2013). A multitude of additive manufacture techniques came into being subsequently in the early 1980s (Jane Bird 2012). Two additional techniques via photo-hardening thermoset polymer or a mask pattern controlling the Ultraviolet exposure area were developed for creation of 3D models with additive manufacturing (Kodama 1981a, b). Bill Master was the first one to herald this era of three-dimensional printing and filed the first-ever patent on

July 2, 1984 (Masters 1987). Subsequently followed by two more which led to the contemporary field today (3-D Printing Steps into the Spotlight 2013).

Stereolithography was discovered by Charles Hull in 1984. In 1986, he elucidated the process in which liquid polymers were solidified under Ultraviolet light to form a three-dimensional model and patented the same (https://www.google.com/patents/us4575330). Electronic data and a computer-aided beam of light were utilized in this method for layerwise production of the model, stacking one layer over the other. In 1988, 3D systems brought about the first commercial SLA printer in the world (Wohlers and Gornet 2014; https://www.google.com/patents/US4041476). By the mid-1990s, novel approaches in this direction were developed at different institutes, e.g., Stanford and Carnegie Mellon University, including spraying and micro casting (Amon et al. 1998; Beck et al. 1992). Constituting and support materials had also reached new heights, conceptualizing and realizing novel geometry forms for objects to be printed (Prinz et al. 1997) (Table 1).

Within the surgical field, 3D printing has been used for the fabrication of representative models or biological scaffolds and also assists in training (Connell and Brown 1997). Three-dimensional imaging data can be converted to standard tessellation file formats for diverse applications (Rosset et al. 2004). Uygun introduced manufacturing of decellularized scaffolds or organs which nucleates live cells (Uygun 2010) hydroxyapatite-based fabricates, for cellular growth support (Petersen 2010; Leukers 2005). Three-dimensional printing is one of the most accepted and constituted in the maxillofacial surgical field where it is used for planning surgery, for fabrication of implants and implant guides (Solar et al. 1992; Mavili et al. 2007; Meehan et al. 2003;

Table 1 Showing all available types of 3D printers (Adapted from https://education.gov.mt/en/resources/News/Documents/Youth%20Guarantee/3D%20Printing.pdf)

Type	Technologies	Materials
Extrusion	Fused deposition modeling (FDM)	Thermoplastics (e.g., PLA, ABS), edible materials, eutectic metals
Granular	Direct metal laser sintering (DMLS)	metal alloys
	Electron beam melting (EBM)	Titanium alloys
	Selective heat sintering (SHS)	Thermoplastic powders
	Selective laser sintering (SLS)	Thermoplastics, ceramic powders, metal powders
	Powder bed and inkjet head 3D printing, Plaster-based	Plaster
Laminated	Laminated object manufacturing (LOM)	Paper, plastic film, metal foil
Light polymerized	Stereolithography (SLA) photopolymer	Photopolymer
	Digital Light Processing (DLP)	Liquid resin

Silva 2008; Flugge 2013). It also used fabrication of anatomical phantom models for training of surgical procedures (Silva 2008).

Use of three-dimensional printing to fabricate surgical instruments is limited. Kondor et al. were the first to develop surgical instrument by using 3D printers (Kondor 2013a, b). In their study for fabrication of general surgical instrument FDM desktop printer is used (Stratasys) (Table 2).

Fused filament fabrication printers are the most common type of consumer-level 3D printer. If the filament is PLA, these scraps may be recycled, but other materials will need to be placed in the trash. Filament bits will adhere to carpet, so smooth flooring or a carpet covering may be preferable. This type of filament can also be

Table 2 Comparison of FFF Filament Material Properties (Adapted from MakerGeeks 2015; MatterHackers Inc. 2015)

Name	Extrusion temperature	Positives	Negatives	Notes
Acrylonitrile butadiene (ABS)	210–250 °C	Strong, smooth surface, flexible, withstands higher temperatures, easy to sand and paint	Strong odor, warps easily, requires heated build plate, may require fume hood, not biodegradable	Same plastic as LEGOs
Polylactic acid (PLA)	190–240 °C	Low warping, recyclable, sweet odor	Less flexible, can be brittle, softens at 60 °C	
High impact polystyrene (HIPS)	230–265 °C	good as support material		Preferred material for LulzBot 3D printers
Polyethylene terephthalate (PET(E/T/G)	210–250 °C	Recyclable, transparent surface, strong		Same plastic as soda bottle, preferred, filament for Cube's Ekocycle
Polyvinyl alcohol (PVA)	190–210 °C	Water soluble	Filament Susceptible to humidity	Useful as support material
Nylon	245 °C	Absorbs color well (see Horne, 2013)	Filament susceptible to humidity	
Ninja Flex	210–240 °C	Prints are flexible	Expensive, adhesion to build plate can be difficult	
Lay Brick and Lay Wood	Varies by intended texture effect	Achieve look of brick or wood using FFF printer	Particles can damage extruder	Mixture of PLA and brick or pine shavings

used to cover the surface of the cabinets and tabletop so that they can be changed frequently and easily cleanable and protect it from contamination (Gonzalez and Bennett 2016).

4 Applications of 3D Printing for Dental Applications

As described in the previous section as per the work practice recommendations from various regulatory bodies worldwide, PPEs assume the mainstay in practicing safe clinical dental practices, in the times of COVID-19 pandemic. Some of the barriers and/or few parts of the PPE kits may be manufactured by the available contemporary 3D printing methods (Tino et al. 2020; Ishack and Lipner 2020; Surface Barriers [internet]. StackPath 2000; Swennen et al. 2020; Canada 2020; https://www.fda.gov/regulatory-information/search-fda-guidance-documents/enforcement-policy-face-masks-and-respirators-during-coronavirus-disease-covid-19-public-health; Amin et al. 2020; https://www.prusa3d.com/covid19/; Coronavirus Disease 2019; Ecker 2020; https://www.makermask.com/; https://3dforcovid.com/; https://www.matterhackers.com/covid-19; https://www.nihlibrary.nih.gov/services/3d-printing-service/3d-printing-medical-equipment-response-covid-19-pandemic; https://www.xyzprinting.com/en-US/material/antibacterial-pla; https://c19hcc.org/static/catalog-resources/health-guidelines-for-3d-printing-safety-equipment.pdf; Burton 2020; Singh and Pandey 2019a, b, 2020a; Ebrahimi and Ju 2018; COVID-19 ACTIVAT 3D 2020; https://bioshield29.com/; https://www.crea3d.com/en/blog/antibacterial-pla-for-medical-applications-b36; https://www.bbc.com/future/article/20200529-the-surfaces-that-kill-bacteria-and-viruses; Kumar et al. 2019; Wei et al. 2015; Singh and Pandey 2019c, 2020b) as given under.

4.1 Face Mask

Face mask is used to protect the person wearing the mask from airborne particles as well as liquid particles from contaminating the face (Tino et al. 2020). N95 respirator masks efficiently filter more than 95% of 0.3-μm airborne particles. Masks should be tested for adequate fit and seal to prevent the entry of air and small droplets from the periphery of the mask (Ishack and Lipner 2020). The Centers for Disease Control and Prevention (CDC) suggests the mandatory usage of N95 masks for healthcare workers during COVID-19 pandemic to prevent cross contamination and infection transmission (Ishack and Lipner 2020). It is based on facial scanning of the person, 3D modeling, and 3D printing. As the new-generation smartphones are coming into play, many applications as well as cameras can be dedicated toward modern imaging techniques like 3D facial scanning. Many organizations including FDA, NIH 3D Print Exchange are working toward the development of N95 masks with different modifications. Numerous others including the researchers, physicians, and commercial companies have proposed designs and are being tested for the degree of protection (Tino et al. 2020).

4.1.1 Methodology

A custom fabricated 3D face mask includes two components which are a face mask and a filter membrane support. It also includes two disposable components which are a head fixation band and a filter membrane. It was created by Swennen and colleagues in April 2020. It involved the following steps mentioned under

(a) **Facial image acquisition**
Scanning of the face is made using a smartphone. These face scans can then be converted to high-definition (HD) data in another objective OBJ file format for further customization.

(b) **3D modeling and design**
Reusable components of the 3D face mask are designed using virtual templates on computer-aided design (CAD) software, the Standard Tessellation Language (STL) (Ishack and Lipner 2020). Components include 3D face mask and the filter membrane support which are connected by screw fixation type to tighten the filter membrane (Fig. 1). To obtain a precise mask fit for the customized masks, a boolean calculation is used to analyze the interaction between the STL file of the face scan and the virtual template of 3D face mask (Ishack and Lipner 2020).
The components to be reused are constructed by a selective laser sintering (SLS) process including a 3D printer and a polyamide composite. Sandblasting is required after it cools down.

(c) **Disposable components**
Two disposable components—a head fixation band and a filter membrane—are made. Velcro bands can be used for head fixation. Disposable filter membrane is developed from polypropylene (PP) and is non-woven melt-blown particle filter membrane (Ishack and Lipner 2020). Disinfection can be done using a solution with broad-spectrum antimicrobial action.

Furthermore, some examples of popular 3D printed face masks are listed and explained below.

Copper3D NanoHack mask—The Copper3D NanoHack mask is printed with a flat piece of polylactic acid (PLA) filament material. It is heated at 55–60 °C (131–140°F) with forced hot air (air dryer) or by putting it in hot water. The mask should be manually checked for an airtight fit seal. Two reusable filters are inserted into masks' simple air intake port. One of the disadvantages of these designs is limited output as a single mask can be printed at a given time. Further, this mask provides a limited amount of airflow when it is sealed and a second breathing port may need to be added (Tino et al. 2020) (Fig. 2).

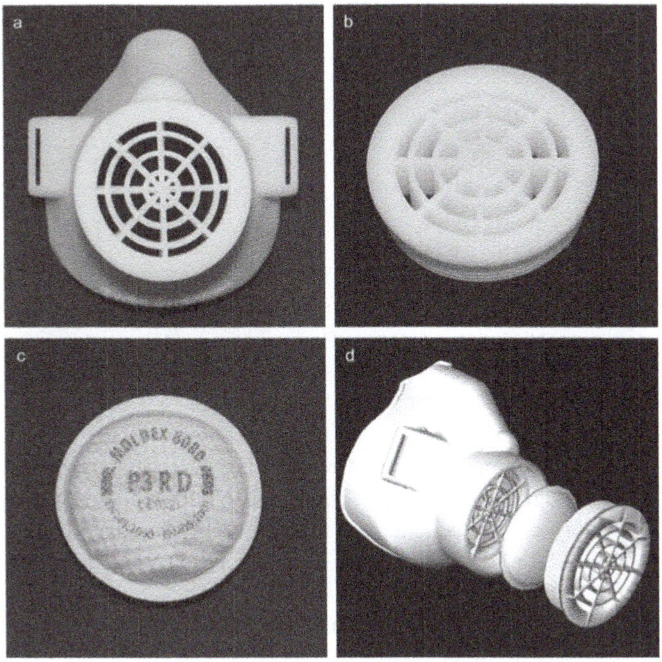

Fig. 1 Custom-made individualized 3D printed protective face mask. *Courtesy* Swennen et al. (2020)

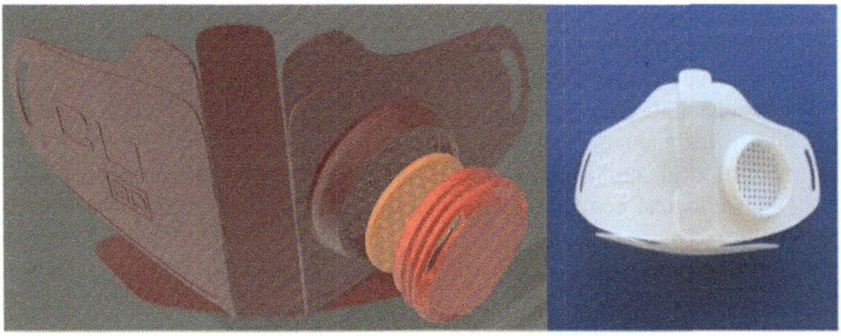

Fig. 2 Copper3D NanoHack mask model. *Courtesy* Tino et al. (2020)

Thingiverse user Kvatthro (HEPA Mask) This mask is made by Polylactic acid filament and manufactured using most desktop printers. It is essential for the mask to have the best possible air seal. Specific masks are present for males and females. Mask contains a port in the front where an exchangeable HEPA filter can be inserted (Tino et al. 2020).

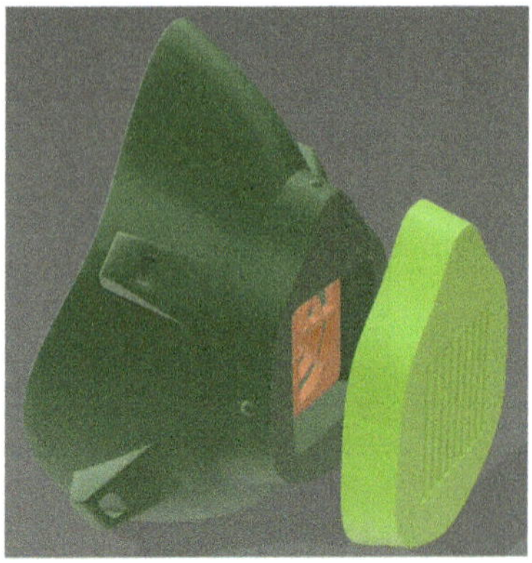

Fig. 3 Lowell Makes mask design. *Courtesy* Tino et al. (2020)

Creality—Chinese company named Creality also suggested a similar design but with a different configuration of the filter holder. The filter holder serves as an insertion platform for layers of folded fabric or filters. Creality goggles can be made from repurposed household plastic. This mask is also pending for seal adequacy check and sizing check (Tino et al. 2020).

Lowell Makes—This mask adopts a different form and serves as the replaceable front filter design. The mask is lined by foam padding on the inside. Lowell Makes masks improve user comfort (Tino et al. 2020) (Fig. 3).

With the advent of different kinds of 3D printed masks, there are few concerns regarding the use.

1. The efficiency of a face mask has to be tested in clinical environment as the performance of the filter membrane, as well as the individual fit around the mouth shall be tested adequately.
2. The polyamide composite material used in the manufacture of masks should be disinfected enough using the disinfectants recommended by the CDC for use against novel coronavirus as this component is reused again and again.
3. Dermatological conditions also should be kept in mind. Allergic lesions or conditions may arise around the nasal bridge where a constant touch occurs for the individual with 3D face mask particularly when worn in warm and humid closed work environments.
4. Virological testing for leakage after disinfection is required before use.

To maintain the standards of workability of these newer developments, there exist some regulations in terms of the materials and practices employed regarding the manufacture of face masks, which need to be adhered to by the developing organizations (Swennen et al. 2020):

ISO 22609 (2004)—Clothing for protection against infectious agents
ASTM F2100 (2019)—Standard Specification for Performance of Materials Used in Medical Face Masks

ASTM F2101 (2019)—Standard Test Method for Evaluating the Bacterial Filtration Efficiency (BFE) of Medical Face Mask Materials, Using a Biological Aerosol of *Staphylococcus aureus*

ASTM F1862/F1862M (2017)—Standard Test Method for Resistance of Medical Face Masks to Penetration by Synthetic Blood (Horizontal Projection of Fixed Volume at a Known Velocity

ASTM F2299 (2003 R2017)—Standard Test Method for Determining the Initial Efficiency of Materials Used in Medical Face Masks to Penetration by Particles Using Latex Spheres.

4.2 Face Shields

CDC guidelines for contact with COVID-19 patients require protective eyewear to be worn at all times. Both N95 mask and a face shield are considered ideal during this time and are accepted by the CDC. A Face shield is a device to shield the eyes and face and has a transparent area (window or visor) in front of the face or visibility (Canada 2020).

4.2.1 Methodology

Face shield can be created using four stages: design, digital preparation, printing, and assembly. A digital design is created in a Standard Tessellation Language (STL) format for constructing a face shield (https://www.fda.gov/regulatory-information/search-fda-guidance-documents/enforcement-policy-face-masks-and-respirators-during-coronavirus-disease-covid-19-public-health).

Next, digital preparation of the STL to a G-code file is done. Specific information regarding the printer parameters (nozzle diameter and position, height limit), print setting (3D printed layer height, horizontal and vertical dimensions), and filament details (type, color, diameter, density) is carried in the G-code file (Fig. 4).

Finally, the shield is assembled using craft materials and office supplies. The velcro is cut into 2–10 inches long strips. The velcro strip holes the frame in place and is

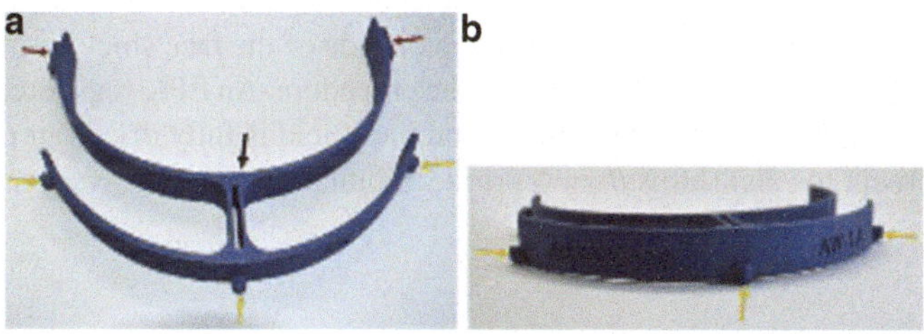

Fig. 4 Face shield: three-dimensionally printed frame. *Courtesy* Amin et al. (2020)

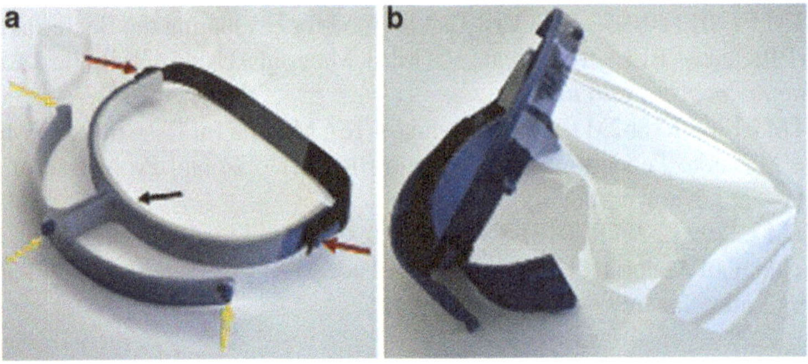

Fig. 5 Assembled face shield. *Courtesy* Amin et al. (2020)

attached on the lateral projections at each end of the frame. A 12-inch long strip of adhesive foam is taken and the sticky side is attached to the inner circumference of the frame. The foam serves as a cushion and prevents mechanical injuries like abrasions, etc., for safe use for long periods. Needful holes are punched approximately, 4 in. apart, on the wide side of the films and each hole is fit snugly into 1 of the 3 rounded projections along the frame (Fig. 5).

The production time depends on a variety of factors such as the printer's default speed, size of the nozzle, extruded layer height, and size of the object. Some 3D printers that can be used are Lulzbot, Ultimaker, Prusa, and large format printers such as the LulzBot TAZ 6 that can print multiple models on one print bed.

Materials

3D printer, polylactic acid material—Velcro strips, adhesive foam, transparency film.

Health Canada advises organizations to consult and follow the standards throughout the design and testing stages:

ANSI/ISEA Z.87.1 (2015)—American National Standard for Occupational and Educational Personal Eye and Face Protection Devices
CSA Z94.3 (2020)—Eye and face protectors
CSA Z94.3.1 (2016)—Guideline for Selection, Use, and Care of Eye and Face Protectors.

According to CDC Guidelines, the disinfection of the face shield is to be done by carefully wiping the inside as well as the outside of the face shield using a clean wipe or cloth with gloved hands only after the procedure. An EPA-registered hospital disinfectant solution can also be utilized and the shield is fully dried (air dry or use clean absorbent towels) (https://www.prusa3d.com/covid19/).

4.3 Use of Disposable Barriers for Protection on Dental Equipment that Can Be 3D Printed in the Near Future

To limit the infection control fluid-impervious barrier on surfaces prone to contamination is placed. Due to pandemic, dental offices have become more aware of the surface barriers that can be used to limit the cross contamination. Barriers are used on light, chair handles, X-ray tubing, chair seat, tabletops (Surface Barriers [internet]. StackPath 2000). Light handles and switches may be touched during the treatment of the individual. Barrier placement films are used for X-ray unit tube heads and control panels. Surface protective covers are available in a variety of shapes and sizes and are cut or pre-fabricated to fit the instrument or equipment they cover (Surface Barriers [internet]. StackPath 2000). It is an adhesive plastic sheet, bags, tubes that sticks or wraps around the surface and is easy to remove at the same time. However, these barriers are not 3D printed for the time being. In near future, it is predicted to be one of the new materials that would need to be 3D printed. Patient bibs, surgical sheets, Sani-sleeve for cords and tubing can also be considered for 3D printing as they are made of plastic-backed paper. Sterilizing pouches of various sizes can also be 3D printed with materials.

5 Materials that Are Antimicrobial and Can Be 3D Printed/Materials and Coatings that Reduce Surface Transmission of Bacteria and Viruses

The surfaces that come in direct contact can be a vector of transmission. The coronavirus can live on cardboard for up to 24 h, and it can stay for up to 3 days on plastic and stainless steel surfaces. Some bacteria such as *Staphylococcus aureus MRSA and Escherichia coli* can live for months on surfaces moreover; infectious yeasts can stay for weeks. Therefore, it is extremely important to disinfect and clean surfaces that are frequently touched. Some scientists also are experimenting with changing the texture of the surfaces and also coating them with substances that kill bacteria and viruses more quickly (Ebrahimi and Ju 2018).

6 Copper

Ancient Indian writings recommend the use of copper water vessels to limit disease (Singh and Pandey 2019a). Water that travels through copper pipes is much less prone to microbial contamination (Singh and Pandey 2019a). In today's times, the antimicrobial nature of copper has been studied and demonstrated extensively.

Spee3D is a manufacturer company from Australia that produces copper and aluminum by 3D printing. Engineer Byron Kennedy used a machine to spray-print a layer of copper onto a door handle (https://c19hcc.org/static/catalog-resources/health-guidelines-for-3d-printing-safety-equipment.pdf). His main aim was to use metal's antiviral and antibacterial properties to counter the threat of the COVID-19 pandemic (https://c19hcc.org/static/catalog-resources/health-guidelines-for-3d-printing-safety-equipment.pdf). Many studies have proven the copper's disinfectant properties as it being antibacterial, antiviral, and antifungal (https://c19hcc.org/static/catalog-resources/health-guidelines-for-3d-printing-safety-equipment.pdf). Spee3D used Melbourne laboratory 360 biolabs to study how copper reacts to SARS-CoV-2 surfaces (https://c19hcc.org/static/catalog-resources/health-guidelines-for-3d-printing-safety-equipment.pdf). The result showed antiviral activity up to 96% and 99.2% in 2 and 5 h, respectively, compared to no such effects on stainless steel surfaces over the same period (Lu et al. 2020). Spee3D customized coat surface sticker made of copper for door handles and push plates (https://c19hcc.org/static/catalog-resources/health-guidelines-for-3d-printing-safety-equipment.pdf). Many healthcare sectors after trial in hospitals including hospitals in Virginia, USA, are using copper as a barrier. However, due to the high cost of the copper, companies and many healthcare companies are not able to adopt the new equipment.

Mechanism—Copper releases free reactive ions known as free radicals as the surface is reactive to environmental influences such as moisture as well as oxygen. These radicals are electrically charged and disrupts the outer membranes of the microbial cell (Burton 2020).

Improvised methods such as combination of ultrasonic-assisted pressure less sintering and 3D printing has been used to fabricate pure copper open cell ordered foams (COCOF) which have a tremendous track record of diverse applications. These foams retain the basic advantages of copper as the parent material and possess better physical characteristics such as thermal and electrical conductivity, sufficient mechanical strength and energy absorption, etc. (Singh and Pandey 2019a, b, 2020a). In fact, metal-filled polymers containing micro-powdered copper have been tried to develop a variety of structures using economic filament extrusion-based 3D printing (Ebrahimi and Ju 2018) (Fig. 6).

Uses

Copper can be used to coat surfaces that are easily contaminable owing to repetitive touch such as door handles, push plates, and railings. This antimicrobial property of copper can be used in hospitals, dental offices, airports, essential workplaces, hotels, and public transport (Burton 2020). Crea3D® collaborated with the manufacturer of antibacterial materials Copper3D, PLACTIVE™ which produces filaments of copper that can be 3D printed. PLACTIVE AN1™ is a nanocomposite developed with a high-quality PLA and highly effective Nano-Copper additive (COVID-19 ACTIVAT 3D 2020). The antibacterial action of PLACTIVE™ has been scientifically tested, and it shows elimination of more than 99.99% of microorganisms (COVID-19 ACTIVAT 3D 2020).

Another American company named BIOSHIELD 29 is manufacturing adhesive sheets of copper for surfaces that are used regularly such as handles, drawer

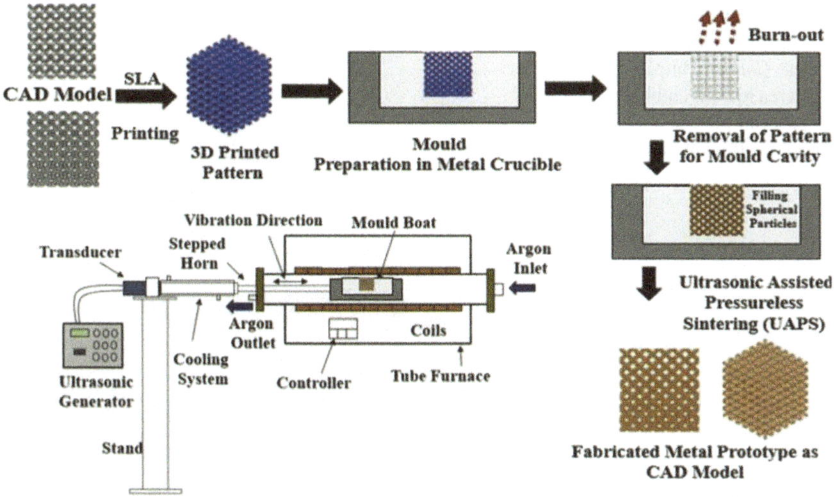

Fig. 6 Combination of ultrasonic-assisted pressureless sintering and 3D printing as a rapid manufacturing technique. *Courtesy* Singh and Pandey (2019a)

openers, and tabletops. The adhesive copper material is water and weather resistant and also is said to be easily removable (https://bioshield29.com/; https://www.crea3d.com/en/blog/antibacterial-pla-for-medical-applications-b36). This will ultimately help reduce the infection. These materials may find wide applications also be used in hospitals and dental offices (Fig. 7).

6.1 Graphene

One such material that is antibacterial and can be 3D printed is "Graphene". Graphene sheets are thin sharp edged. It can cut through the bacterial membrane. Graphene disrupts the membrane integrity of the bacterial cells also as another means to kill the bacterial cells. The material acts as an electron acceptor and interacts in the same capacity in the cellular processes occurring in from bacterial membranes, leading to changes in the bacterial cell membrane permeability and leaking of cellular contents (Kumar et al. 2019). In addition, graphene is renewable, mass producible, and cost effective as compared to other nanomaterials. Such graphene-based nanomaterials are being used to develop 3D printed nanocomposites which have high antibacterial properties (Wei et al. 2015). And can be useful additions for antibacterial surfaces for dental and medical offices to prevent cross contamination and infection.

Further, combination materials such as a new class of topological ordered copper graphene foams (TOCGF) are produced by 3D printing coupled with ultrasonic-assisted pressureless sintering. The relative density and compressive properties of the

Fig. 7 BioShield 29 antimicrobial adhesive copper. *Courtesy* https://www.crea3d.com/en/blog/antibacterial-pla-for-medical-applications-b36

TOCGF were improved with respect to pure copper with increasing the concentration of graphene up to 1% but decreased with further addition (Singh and Pandey 2019, 2020).

6.2 *Polylactic Acid (PLA)*

Company named XYZ Printing is printing filaments of antibacterial polylactic acid (PLA) that includes silver ions to inhibit bacterial growth by 99% and helps in reducing the risk of infections (https://www.xyzprinting.com/en-US/material/antibacterial-pla). Its antibacterial agent also follows The Restriction of Hazardous Substances directive (RoHS) regulations which makes it the ideal 3D printing material for households and schools (https://www.xyzprinting.com/en-US/material/antibacterial-pla) from a dental point of view, X-ray holders can be made of antibacterial PLA filaments. Tubing and valves used for suction can also be made with PLA. Chair handles, light handles, and many other parts of the dental chair can also be manufactured with the help of PLA.

7 Evolution and Innovations in 3D Printing Technologies to Combat COVID-19 Safe Dentistry

Dental professionals, including lab technicians, dental product manufacturers, hygienists, are finding ways to put their skills and equipment to use by using their 3D printers for the production of PPE. Moreover, faculty members and students from dental schools are using 3D printers and stereolithography and fused deposition modeling to produce reusable visors.

Professional additive manufacturers and designers in the 3D printing community are volunteering to manufacture PPE and are also providing free software online to individuals who have a 3D printer at home that can contribute to producing PPE.

iCREATE Face Shields by David Ecker. He is working on innovation at Stony Brook University that can help the society during a pandemic (Coronavirus Disease 2019). Materials advised to make a face shield are 3D Printed Raft out (it's a u-shaped object), half-moon foam shape for the rim, Elastic to 10″ in length.

- Maker Mask—NIH approved 3D printable respirator mask that uses local material. HEPA Filter respirator is easily replaceable in the mask (Ecker 2020).
- 3D for COVID—volunteers helping make PPE for first-responders treating COVID-19 (https://www.makermask.com/).
- Matter Hackers COVID-19 Maker Response Hub.

They help connect those in need of medical aid such as hospitals and those who can create it using digital manufacturing (https://3dforcovid.com/).

7.1 Public-Private Partnerships a Way Forward to Coordinate 3D Healthcare Products for the COVID-19 Response

The Food and Drug Administration (FDA), the Veterans Health Administration (VA), the National Institutes of Health (NIH), and America Makes are partnering to make PPE available through 3D printing (https://www.matterhackers.com/covid-19).

America Makes is in the process of devising an online repository for arenas in need of medical PPE, developers capable of providing 3D printing facilities, and designers who can devise 3D printing amenable designs. Moreover, they are also connecting manufacturers with printable designs being reviewed for clinical use to ensure the delivery of safe PPE for medical caregivers on the front-line of the COVID-19 pandemic (https://www.nihlibrary.nih.gov/services/3d-printing-service/3d-printing-medical-equipment-response-covid-19-pandemic).

8 Critical Issues Pertaining to Applications of 3D Printing for Clinical Dental Practice

Global healthcare systems are under severe shortage of PPE due to pandemic. With the urgent need for equipment due to the growing COVID-19 crisis, 3D printing is being used. Standard protocols for safety and quality checks of 3D printing labs should be followed. Many of the PPE designs including face masks and face shields are a work in progress, and the quality of these products must be evaluated carefully. Moreover, most of the equipment is for reuse, therefore, correct sterilization and disinfecting techniques should be considered and safety protocols should always be reviewed. Despite the good intent behind most 3D printing, there are some concerns which need a serious scientific deliberation, before implementation or commercialization.

8.1 3D Printed PPE and Other Parts Vis a Vis FDA-Approved PPE and Medical Devices

FDA's most recent advice relevant to N95 respirators and face shields is as under

- 3D printed personal protective equipment (PPE) does not provide a similar degree of protection as a fluid barrier and air filtration as FDA-cleared surgical masks and N95 respirators.
- 3D printed FDM parts are porous which creates risk of infectious material in 3D printed PPE.
- 3D printed masks should be checked for leaks, breathability through the filter, caution should be exercised in a surgical environment with liquid barrier protection and flammability concerns.
- Verification of 3D printed products fit and work efficiently should be checked properly before use in a clinical setting.

8.2 Sterilization and Sanitization During Manufacture and Delivery of 3D Printed Parts

It is possible that 3D printed parts of PPE may not be compatible with the sterilization techniques hospitals and clinics currently use. As it is known that SARS-CoV-2 virus can survive two to three days on plastic surfaces, it is important to note that an individual who is infected with the virus can transfer the virus to someone else via a 3D printed product. PLA, the most common 3D printing material, isn't durable enough to withstand the high heat and chemicals used for sterilization. It is also impossible to guarantee how sterilized these products are because of where they are made and packaged.

Most Maker labs are not confined to FDA-approved levels for sterility in producing medical-grade items, additional measures should be followed for sanitization prior to supporting the COVID-19 response. A proper protocol needs to be adhered to the manufacture of such 3D printing appliances.

- Sanitize the area: In case of COVID-19 exposure of anybody sharing the space in which 3D printing has been carried out, the work needs to be halted for at least 2 weeks. Complete sanitization of all areas including material storage, preparation, and for packaging for onward transportation of finished parts shall be ensured.

Thus, extra time should be taken to make sure that the sterilization of these products is apt and meets the standards of optimum sterilization.

8.3 Production Time

3D printed face shields and masks usually take more than an hour to print on a standard desktop 3D printer. However, factories produce masks in seconds using the injection molding technique.

9 Summary and Conclusions

COVID-19 pandemic has increased the need for Personal Protective Equipment and medical equipment. Healthcare professionals are at high risk due to the shortage of PPE all over the world. Fortunately, the technology of 3D printing shows a tremendous potential to overcome the lack of these equipments. The 3D printing community continues to develop a variety of reusable PPE products by the use of low-cost 3D printers which include face masks (N95 and surgical), face shields, disposable barriers which cover the surfaces prone to the deposition of virus, metal coatings which use copper and dental materials such as nanocomposites with the addition of graphene. The manufacture of 3D printed copper sheets and graphene nanocomposites is relatively uncommon but is expected to increase in the future. During this pandemic, an organized communication between the 3D printing community and the hospital chain should be established to ensure planned production and distribution.

Despite the state of this growing crisis, safety of the produced healthcare equipment is of utmost importance. There are chances that a 3D printing PPE manufacturing protocol may be developed in the coming future, and it is also necessary to develop policies and funding mechanisms for these projects. The challenges are many and one of the challenges is keeping a suitable level of sterility of the manufactured equipment. Another one is that the efficiency of a face mask needs to be tested in clinical environment. The performance of the filter membrane, as well as the individual fit around the mouth, must be tested adequately. Furthermore, it is possible that 3D printed parts of PPE may not be compatible with the sterilization techniques hospitals and clinics currently use and there is a need to modify these techniques. Mass production of equipment might be limited due to the use of 3D printers which are non-industrial or in-office printers.

Self-sustaining production of face masks and shields using one's own printing capabilities in combination with sufficiently available commercial goods can prove to be a great aid to help protect the healthcare workers. In general, the 3D printing technology will result in increased flexibility, independence, and autonomy in emergency situations where the requirement of PPE might increase (Wesemann 1997).

The devices for health care including dentistry tools and equipment shall be highly regulated for safety. The 3D printing industry must work in parallel to ensure that 3D parts manufactured during pandemic times are safe or at least a minimum safer than the alternative of not using them during a pandemic. Standard safety and quality measures of 3D printing labs should be in place to be followed, even with the growing urgency of the COVID-19 crisis. Partnerships between university-based 3D printing resources and hospitals along with the upkeep of appropriate safety protocols should be emphasized. It is imperative for the national agencies and governments to take the charge of rebuilding the nation by adopting newer cost-effective techniques such as 3D printing technologies to cater to basic needs and equipment to maintain safe dental practices worldwide. 3D printing techniques shall provide ways and means to strengthen and sustain "Dentistry" as a profession by targeted policy-makers and material capacity building for safe and effective oral healthcare practices post-pandemic.

References

Amanda SU, Al'Aref SJ (2018) History of 3D printing. In: Al'Aref SJ (ed) 3D printing applications in cardiovascular medicine, 1st edn. Academic press in an imprint of Elsevier, Cambridge, p 01

Amin D, Nguyen N, Roser SM, Abramowicz S (2020) 3D printing of face shields during COVID-19 pandemic: a technical note [Internet]. https://www.ncbi.nlm.nih.gov/pmc/articles/PMC7194067/. Accessed 24 Sep 2020

Amon CH, Beuth JL, Weiss LE, Merz R, Prinz FB (1998) Shape deposition manufacturing with microcasting: processing, thermal and mechanical issues. J Manuf Sci Eng 120:656–665

Bachtiar EO, Erol O, Millrod M et al (2020) 3D printing and characterization of a soft and biostable elastomer with high flexibility and strength for biomedical applications. J Mech Beh of Biomed Materials 104:103649

Banerjee SS, Burbine S, Kodihalli N et al (2019) 3D-printable pp/sebs thermoplastic elastomeric blends: preparation and properties. Polymers (Basel) 11:347

Barzon L, Pacenti M, Berto A, Sinigaglia A, Franchin E, Lavezzo E et al (2016) Isolation of infectious Zika virus from saliva and prolonged viral RNA shedding in a traveller returning from the dominican Republic to Italy. Eurosurveillance 21:30159

Beaman JJ, Barlow JW, Bourell DL, Crawford RH, Marcus HL, McAlea KP (2013) Solid freeform fabrication: a new direction in manufacturing: with research and applications in thermal laser processing [internet]. Illustrated. Springer Science & Business Media. https://books.google.com/books?Id=IPvxBwAAQBAJ&pg=PA47&lpg=PA47&dq=swainson+patent&source=bl&ots=xxIkDFBwUD&sig=I_ErMVsAyundvQ07g3-BaPt3yJs&hl=en&sa=X&ved=0ahUKEwiMp8rL76bUAhVH_IMKHbngAXIQ6AEIUDAM#v=onepage&q=swainson%20patent&f=false

Beck JE, Fritz B, Siewiorek D, Weiss L (1992) Manufacturing mechatronics using thermal spray shape deposition. In: Proceedings of the 1992 solid freeform fabrication symposium. Accessed on 20 Dec 2014

Belser JA, Rota PA, Tumpey TM (2013) Ocular tropism of respiratory viruses. Microbiol Mol Biol Rev 77:144–56

Burton M (2020) Copper takes aim at COVID-19 with virus-killer coatings [Internet]. Reuters. Thomson Reuters. https://www.reuters.com/article/us-health-coronavirus-copper-antimicrobi/copper-takes-aim-at-covid-19-with-virus-killer-coatings-idUSKBN22K0RX

Cai M, Li H, Shen S et al (2018) Customized design and 3D printing of face seal for an N95 fifiltering facepiece respirator. J Occup Environ Hyg 15:226–34

Canada H (2020) Government of Canada [Internet]. COVID-19: personal protective equipment 3D printing and manufacturing—Canada.ca./Government du Canada. https://www.canada.ca/en/health-canada/services/drugs-health-products/medical-devices/covid-19-unconventional-manufacturing-personal-protective-equipment.html. Accessed 24 Sep 2020

Center for Devices and Radiological Health. Face Mask and Respirator Enforcement Policy During COVID-19—Guidance [Internet]. U.S. Food and Drug Administration. FDA. https://www.fda.gov/regulatory-information/search-fda-guidance-documents/enforcement-policy-face-masks-and-respirators-during-coronavirus-disease-covid-19-public-health. Accessed 24 Sep 2020

Connecting makers across the country to provide critical medical supplies for doctors in need. 3D Printing for coronavirus. https://3dforcovid.com/. Accessed 24 Sep 2020

Connell Berry, Brown Craven (1997) Preliminary experience with medical applications of rapid prototyping by selective laser sintering. Med Eng Phys 19:90–6

Coronavirus Disease 2019 (COVID-19) [Internet]. centers for disease control and prevention. centers for disease control and prevention. https://www.cdc.gov/coronavirus/2019-nCoV/index.html. Accessed 24 Sep 2020

COVID-19 ACTIVAT 3D Copper (2020) SPEE3D. https://www.spee3d.com/activat3d-copper/. Accessed 24 Sep 2020

COVID-19 coronavirus pandemic (2020). https://www.worldometers.info/coronavirus/. Accessed on 11 Aug 2020

COVID-19 Healthcare Coalition. https://c19hcc.org/static/catalog-resources/health-guidelines-for-3d-printing-safety-equipment.pd. Accessed 24 Sep 2020

COVID-19symptoms may include altered senses of smell, taste [internet] (2020). https://abcnews.go.com/Health/covid-19- symptoms-include-altered-senses-smell-taste/story?id = 69769629

del RC, Malani PN (2020) 2019 novel coronavirus—important information for clinicians. JAMA 323:1039

Ebrahimi ND, Ju YS (2018) Thermal conductivity of sintered copper samples prepared using 3D printing-compatible polymer composite filaments. Addit Manuf 24:479–485

Ecker D (2020) Information Technology professional is the Director of iCREATE at Stony Brook University, Ecker AD, David Ecker is Information Technology professional is the Director of iCREATE at Stony Brook University, Www.nyinnovate.com, et al. Face Shields iCREATE [Internet]. NY Innovate. https://nyinnovate.com/2020/03/26/face-shields-icreate/. Accessed 24 Sep 2020

Fan Y, Zhao K, Shi ZL, Zhou P (2019) Bat Coronaviruses in China. Viruses 11:210

Fehr AR, Perlman S (2015) Coronaviruses: an overview of their replication and pathogenesis. In: Coronaviruses. Humana Press, New York, NY, pp 1–23

Flugge TV (2013). Three-dimensional plotting and printing of an implant drilling guide: simplifying guided implant surgery 71:1340–1346

Gonzalez SAR, Bennett DB (2016) 3D printing and printers. In: Russell Gonzalez. 3D printing: a practical guide for librarians, no 22, 1st edn. Rowman &Littlefield, Lanham, pp 11–30

Home [Internet]. Crea3D®. https://www.crea3d.com/en/blog/antibacterial-pla-for-medical-applications-b36. Accessed 24 Sep 2020

Home [Internet]. Maker Mask. https://www.makermask.com/. Accessed 24 Sep 2020

https://education.gov.mt/en/resources/News/Documents/Youth%20Guarantee/3D%20Printing.pdf. Accessed on 10 June 2020

https://www.cdc.gov/coronavirus/2019-ncov/hcp/dental-settings.html. Accessed on 10 June 2020

https://www.osha.gov/Publications/OSHA4019.pdf Accessed on 10 June 2020

Huang C, Wang Y, Li X, Ren L, Zhao J, Hu Y et al (2020) Clinical features of patients infected with 2019 novel coronavirus in Wuhan China. Lancet 395:497–506

Ishack S, Lipner SR (2020) A review of 3-dimensional skin bioprinting techniques: applications, approaches, and trends [e-pub ahead of print]. Dermatol Surg

Ishack S, Lipner SR (2020) Applications of 3D printing technology to address COVID-19-related supply shortages [Internet]. https://www.ncbi.nlm.nih.gov/pmc/articles/PMC7172844/. Accessed 24 Sep 2020

Jane Bird (2012) Exploring the 3D printing opportunity. Financial Times. Accessed on 30 Aug 2012

Kodama H (1981a) A scheme for three-dimensional display by automatic fabrication of three-dimensional model. IEICE Trans Electron (Jpn Ed) 4:237–41

Kodama H (1981b) Automatic method for fabricating a three-dimensional plastic model with photo-hardening polymer. Rev Sci Instrum 52:1770–73

Kondor S (2013a) On demand additive manufacturing of a basic surgical kit. J Med Dev ASME 7

Kondor S (2013b) Personalized surgical instruments. J Med Dev ASME 7

Kumar P, Huo P, Zhang R, Liu B (2019) Antibacterial properties of graphene-based nanomaterials. Nanomaterials 9(5):737

Leukers B (2005) Hydroxyapatite scaffolds for bone tissue engineering made by 3D printing. J Mater Sci Mater Med 16:1121–1124

Lu B, Huang Y, Huang L, Li B, Zheng Z, Chen Z et al (2010) Effffect of mucosal and systemic immunization with virus-like particles of severe acute respiratory syndrome coronavirus in mice. Immunology 130:254–61

Lu CW, Liu XF, Jia ZF (2020) 2019-nCoV transmission through the ocular surface must not be ignored. Lancet 395:e39

MakerGeeks (2015) PLA filament, ABS filament & 3D printing supplies at markergeeks. Com. MakerGeeks.com. http://www.makergeeks.com/

Marco C, Rajnik M, Cuomo A, et al (2020) Features, evaluation and treatment coronavirus (COVID-19). Treasure Island FL: Stat Pearls [Internet]

Masters WE (1987) United States Patent: 4665492—Computer automated manufacturing process and system

MatterHackers. COVID-19 additive manufacturing response—USA. MatterHackers. https://www.matterhackers.com/covid-19. Accessed 24 Sep 2020

MatterHackers Inc. (2015) "Home". MatterHackers Inc. http://www.matterhackers.com

Mavili ME, Canter HI, Saglam-Aydinatay B, Kamaci S, Kocadereli I (2007) Use of three- dimensional medical modeling methods for precise planning of orthognathic surgery. J Cranio-fac Surg 18:740–747

Meehan M, Teschner M, Girod S (2003) Three-dimensional simulation and prediction of craniofacial surgery. Orthod Cranio-fac Res 6(supplement 1):102–107

Meng L, Hua F, Bian Z (2020) Coronavirus disease 2019 (COVID-19): emerging and future challenges for dental and oral medicine. J Dent Res 12

Patent US 4,041,476-Method, medium and apparatus for producing three-dimensional figure product—Google patents [internet]. https://www.google.com/patents/US4041476

Patent US 4,575,330-Apparatus for production of three-dimensional objects by stereolithography-Google patents [Internet]. https://www.google.com/patents/us4575330

Peng X, Xu X, Li Y, Cheng L, Zhou X, Ren B (2020) Transmission routes of 2019-nCoV andcontrols in dental practice. Int J Oral Sci 12:1–6

Perlman S, Netland J (2009) Coronaviruses post-SARS: update on replication and pathogenesis. Nat Rev Microbiol 7:439–50

Petersen TH (2010) Tissue-engineered lungs for in vivo implantation. Science 329:538–541

Prinz FB, Merz R, Weiss L, Ikawa N (ed) (1997) Building parts You Could Not build before. In: Proceedings of the 8th international conference on production engineering. Chapman & Hall, London, UK, pp 40–44

Protect Areas From Viral, Fungal, and Bacteria [Internet]. Bioshield29. https://bioshield29.com/. Accessed 24 Sep 2020

Ranney ML, Griffeth V, Jha AK (2020) Critical supply shortages—The need for ventilators and personal protective equipment during the COVID-19 pandemic [e-pub head of print]. N Engl J Med

Rosset A, Spadola L, Ratib O (2004) OsiriX: an open-source software for navigating in multidimensional DICOM images. J Digit Imaging 17:205–216

Rothe C, Schunk M, Sothmann P, Bretzel G, Froeschl G, Wallrauch C, et al (2020) Transmission of 2019-nCoV infection from an asymptomatic contact in Germany. N Engl J Med

Sabino-Silva R, Jardim AC, Siqueira WL (2020) Coronavirus COVID-19impacts to dentistry and potential salivary diagnosis. Clin Oral Invest 20:1–3

Segal A, Wong DT (2008) Salivary diagnostics: enhancing disease detection and making medicine better. Europ J Dental Educ: Off J Assoc Dental Educ Eur 12(Suppl 1):22

Silva DN (2008) Dimensional error in selective laser sintering and 3D-printing of models for craniomaxillary anatomy reconstruction. J Craniomaxillofac Surg 36:443–9

Singh G, Pandey P (2019a) Uniform and graded copper open cell ordered foams fabricated by rapid manufacturing: surface morphology, mechanical properties and energy absorption capacity. Mater Sci Eng A 761:138035. https://doi.org/10.1016/j.msea.2019.138035

Singh G, Pandey P (2019b) Rapid manufacturing of copper components using 3D printing and ultrasonic assisted pressureless sintering: experimental investigations and process optimization. J Manuf Process 43:253–269. https://doi.org/10.1016/j.jmapro.2019.05.010

Singh G, Pandey PM (2019c) Topological ordered copper graphene composite foam: fabrication and compression properties study. Mater Lett 257:126712

Singh G, Pandey PM (2020a) Experimental investigations into mechanical and thermal properties of rapid manufactured copper parts. Proc Inst Mech Eng Part C: J Mech Eng Sci 234(1):82–95. https://doi.org/10.1177/0954406219875483

Singh G, Pandey PM (2020b) Rapid manufacturing of copper-graphene composites using a novel rapid tooling technique. Rapid Prototyp J

Solar P, Ulm C, Imhof H et al (1992) Precision of threedimensional CT-assisted model production in the maxillofacial area. Eur J Radiol 2:473–477

Surface Barriers [internet]. StackPath (2000). https://www.dentaleconomics.com/practice/article/16388943/surface-barriers. Accessed 24 Sep 2020

Swennen GRJ, Pottel L, Haers PE (2020) Custom-made 3D-printed face masks in case of pandemic crisis situations with a lack of commercially available FFP2/3 masks [Internet]. Int J Oral Maxillofac Surg. Published by Elsevier Ltd on behalf of International Association of Oral and Maxillofacial Surgeons. https://www.ncbi.nlm.nih.gov/pmc/articles/PMC7132499/. Accessed 24 Sep 2020

The surfaces that kill bacteria and viruses [Internet]. BBC Future. BBC. https://www.bbc.com/future/article/20200529-the-surfaces-that-kill-bacteria-and-viruses. Accessed 24 Sep 2020

Tino R, Moore R, Antoline S, Ravi P, Wake N, Ionita CN, et al (2020) COVID-19 and the role of 3D printing in medicine. 3D Print Med 6(1)

To KK, Tsang OT, Chik-Yan Yip C, Chan KH, Wu TC, Chan JMC, Leung WS, Chik TS, Choi CY, Kandamby DH et al (2020) Consistent detection of 2019 novel coronavirus in saliva. Clin Infect Dis [epub ahead of print 12 Feb 2020]; pii: ciaa149

To KK, Tsang OT, Yip CC, Chan KH, Wu TC, Chan JM, et al (2020) Consistent detection of 2019 novel coronavirus in saliva. Clin Infect Dis

3D printed face shields for medics and professionals [Internet]. Prusa3D.https://www.prusa3d.com/covid19/. Accessed 24 Sep 2020

3D Printing Medical Equipment in Response to the COVID-19 Pandemic [Internet]. National Institutes of Health. U.S. Department of Health and Human Services. https://www.nihlibrary.nih.

gov/services/3d-printing-service/3d-printing-medical-equipment-response-covid-19-pandemic. Accessed 24 Sep 2020
3-D Printing Steps into the Spotlight (2013) Upstate Bus J. Accessed on 20 Dec 2019
Uygun BE (2010) Organ reengineering through development of a transplantable recellularized liver graft using decellularized liver matrix. Nat Med 16:814–820
Wax RS, Christian MD (2020) Practical recommendations for critical care and anesthesiology teams caring for novel coronavirus (2019-nCoV) patients. Canad J Anesth/J Canad d'anesthésie 12:1–9
Wei X, Li D, Jiang W, Gu Z, Wang X, Zhang Z, et al (2015) 3D printable graphene composite. Sci Rep 5(1)
Weiss SR, Leibowitz JL (2011) Coronavirus pathogenesis. Adv Virus Res (Academic Press) 81:85–164
Wesemann C et al (1997) 3-D printed protective equipment during COVID-19 pandemic. Materials (Basel, Switzerland) 13(8). https://doi.org/10.3390/ma13081997
Wohlers T, Gornet T (2014) History of additive manufacturing. Wohlers Associates, Inc.
Wu F, Zhao S, Yu B (2020) A new coronavirus associated with human respiratory disease in China. Nature 579:265–9
Xie X, Li Y, Sun H, Liu L (2009) Exhaled droplets due to talking and coughing. J R Soc Interface 6(suppl_6):S703–S714
Xu J, Li Y, Gan F, Du Y, Yao Y (2020a) Salivary glands: potential reservoirs for COVID-19 asymptomatic infection. J Dent Res. https://doi.org/10.1177/0022034520918518
Xu H, Zhong L, Deng J, Peng J, Dan H, Zeng X, Li T, Chen Q (2020b) High expression of ACE2 receptor of 2019-nCoV on the epithelial cells of oral mucosa. Int J Oral Sci 12:8
Xu Z, Shi L, Wang Y, Zhang J, Huang L, Zhang C, et al (2020) Pathological findings of COVID-19 associated with acute respiratory distress syndrome. Lancet Respiratory Med 17
xyzprinting.com. https://www.xyzprinting.com/en-US/material/antibacterial-pla. Accessed 24 Sep 2020
Yin Y, Wunderink RG (2018) MERS, SARS and other coronaviruses as causes of pneumonia. Respirology 23:130–7
Yu I, Chen RK (2020) A feasibility study of an extrusion-based fabrication process for personalized drugs. J Pers Med 10:16
Zhou P, Yang XL, Wang XG, Hu B, Zhang L, Zhang W et al (2020) Discovery of a novel coronavirus associated with the recent pneumonia outbreak in humans and its potential bat origin. BioRxiv 1:1–18
Zuanazzi D, Arts EJ, Jorge PK, Mulyar Y, Gibson R, Xiao Y, Siqueira WL, et al (2017) Postnatal identifification of Zika virus peptides from saliva. J Dent Res 96:1078–1084

Developing Personal Protective Equipment Against COVID-19

Sagarika Bhattacharjee and Harmanpreet Singh

Abstract In this current situation of the COVID-19 pandemic, a huge loss of human lives has been faced by the world. The situation seems to be uncontrollable both because of the widespread of the virus and contamination of surroundings due to it, as well as a longer period of lockdown has emerged new social and economic problems. So, it is advised by the experts that this situation has to be dealt with by taking care of many precautions in mind while carrying out our daily life routine. On the other hand, frontline workers, policemen, and doctors need to protect themselves even after being exposed to the affected patients. For their safety and others, personal protective equipment (PPE) is the safeguard in this situation. The prime aim of PPE is to break the chain and avoid human to human contact. To develop these PPEs, scientists and experts have been designing and manufacturing kits consisting of various types of masks, face visors, gloves, glasses, shoe covers, surgical suits, and caps. Keeping the dimensions of the novel coronavirus in mind, i.e., 80–220 nm, the PPE kits are made to allow ventilation to the body along with the sole purpose of protection. For this, various methods of manufacturing these types of equipment are adapted, especially 3D printing is widely used for making face visors. The potential applications of nanomaterials came into the picture due to their properties such as antibacterial, hydrophobicity, transparency, and flexibility. Here, the chapter discusses the development of such PPEs for the protection of frontline workers and doctors along with the consequences followed by their usage.

Keywords Healthcare · COVID-19 · 3D printing · Personal protective equipment

---

S. Bhattacharjee (✉)
Department of Physics and Nanotechnology, SRM Institute of Science and Technology (Deemed to be University), Kattankulathur 603203, Tamil Nadu, India

H. Singh
Department of Mechanical Engineering, Thapar Institute of Engineering and Technology, Patiala 147004, Punjab, India

K. Sandhu et al. (eds.), *Emerging Applications of 3D Printing During CoVID 19 Pandemic*, Lecture Notes in Bioengineering, https://doi.org/10.1007/978-981-33-6703-6_8

Abbreviations

COVID	Coronavirus
WHO	World Health Organization
PPE	Personal Protective Equipment
3D	Three Dimensional
VHA	Veterans Health Administration
3DP	Three Dimensional Printing
PLA	Poly Lactic Acid
CDC	Centers of Disease Control
FDM	Fused Deposition Modeling
LS	Laser Sintering
SLS	Selective Laser Sintering
PVC	PolyVinyl Chloride
ABS	Acrylonitrile Butadiene Styrene
DMLS	Direct Metal Laser Sintering

1 Introduction

The coronavirus, commonly called COVID-19, has really brought up a whole lot of trouble and finally ended up as a pandemic. The deadly virus, yet with very common symptoms like common flu, has spread so quickly and created a lot of chaos. As reported by the World Health Organization (WHO) till 23rd November 2020, the total number of confirmed cases went as high as 58,425,681 among which the number of deaths reported was 1,385,218 (World Health Organization 2020). Hence, WHO declared it as "Public-health emergency" by the end of January 2020 (Li et al. 2020). This situation has put forward the true fragility of the conventional global supply chains (Kumar and Chandra 2010; Hoffman 2006). The healthcare sector has been greatly challenged in this awful time. To deal with this, numerous efforts have been taken to take care of the health of the frontline workers, i.e., to save more lives and serve the infected people, doctors, and various other frontline workers who need the utmost care and their own safety. Along with the medical and nonmedical equipment facing a huge shortage, personal protective equipment (PPE) kits for healthcare and frontline workers, low availability of ventilators for patients, an inadequate amount of sanitizer solutions availability, and an insufficient number of coronavirus test kits for the general public has also been observed due to the fragility of supply chains. Countries stopped exporting PPE kits to supply enough to their own people. To overcome the shortage, scientists and entrepreneurs tried their best to develop new technologies and fetch other alternatives to meet the demand to the earliest and save more lives who serve the nation.

For the past several years, 3D printing (additive manufacturing) has greatly contributed to the medical sector. For instance, surgical implants (Pettersson et al.

2020; Akmal et al. 2018; Zhang et al. 2020; Matai et al. 2020), robotic arms (Robotic Surgery: Custom 3D Printed Surgical Instruments for Every Patient 2020), biological tissues (Wang et al. 2008; Freeman et al. 2019), and organs, also some other equipment and tools have been manufactured by 3D printing. The main advantages of using 3D printing in this sector include dimensional customization, high accuracy, reproducibility, availability on-demand with wide options in terms of designs. Not only that, but 3D printing also provides digital designs, faster delivery time, and flexibility in the processing of any part of the design which is not available in the case of conventional methods. In the present COVID-19 pandemic, 3D printing has been rapidly used to print various equipment and parts for medical devices and protective gear, due to the lack of resources at the time of such an emergency. According to the search volume data of Google, the term "3D printing" was searched 2.2 times more in April 2020 than in the same period last year (Salmi et al. 2020). In this present chapter, the focus is to understand the importance of 3D printing to overcome the shortage of medical supplies specifically PPE kits, and the advancements that happened during this time.

2 Global Scenario of Additive Manufacturing-Based Industries

As it has been seen in the current situation of crisis, the prime importance is to protect human lives. The most viable solution is to frequently sanitize ourselves, as well as our surroundings, also isolate ourselves as much as possible. Basically, for frontline workers and doctors, it's difficult. So, the best way is to provide them with the best PPE kits. Looking at the global scenario, the quick solution for the shortage in PPE kits can be solved with the help of additive manufacturing industries. Some of the examples are stated below in Table 1.

This is how 3D printing has helped in keeping the supply chain intact. With the increase in demand, the supply can be subsequently increased to keep up with requirements. Not only that, hygiene can be maintained while using 3D printing techniques.

3 3D Printing Involved in PPE Production

PPE mainly consists of various components such as gowns, gloves, goggles, masks, head covers, face shields, and shoe covers. The importance of each component is discussed in brief in Table 2.

Out of the abovediscussed components, 3D printed parts are limited to face shields, ventilators in masks, and some transparent gowns and headcovers. Still, there is a lot of scope for other products to be manufactured.

Table 1 3D printing companies respond to COVID-19 (Examplatory use of 3D printing to provide medical supplies during coronavirus (COVID-19) Pandemic in 2020)

Names of 3D printing company	No. of parts produced
Consortium—Formlabs, Carbon, Envision Tec, and Origin: nasopharyngeal swabs (potential weekly capacity)	40,00,000
Nexa3D (3D Printing manufacturer, U.S.): Test swabs (potential weekly production capacity)	5,00,000
Stratasys and Origin (U.S.): Nasopharyngeal swabs (Production capacity per day)	1,90,000
Nissan (Car manufacturer, Japan): Face shields (Potential weekly production capacity)	1,00,000
The Voodoo manufacturing (3D printing unit, U.S.): face shields and swabs (Weekly capacity for 2,500 face shields and 50,000 swabs)	52,500
Ricoh 3D (Printing, UK): Face shields (Weekly capacity)	40,000
3D Hubs (3D manufacturing unit, Netherlands): Face shields (Through the COVID-19 Manufacturing Fund)	20,000
The Forecast 3D (Industrial 3D printing, U.S.): Face shields, stopgap masks, nasopharyngeal swabs, and other PPE products (daily part production capacity)	10,000
The Nexa3D (3D printing manufacturer, U.S.): Face shields (potential weekly production capacity)	10,000
Prusa Research (3D printing company, Czech Republic): Face shields	10,000
Mobility/Medical goes Additive consortium (nearly 50 enterprises, Germany): Face shields	5,000
PERA CD—N95 mask lining bracket - Farsoon Technologies (China)—Safety goggle and Mask adjuster (Large scale Safety googles PPE manufacturer, China)	2,000
The Protolabs (3D Printing company): ventilator components	3,000
Fast Radius (Additive manufacturing solutions, U.S.): face shield kits (the potential daily production capacity of 10,000)	1,500
Azul3D (3D Printing manufacturer): face shields (Current daily capacity)	1,000
Smile direct club (Digital Dentistry enterprise): face shields (potential capacity of 7,500 per day)	1,000
Photocentric (3D printing company, UK): valves for respirators (trial run; potential capacity of 40,000 per week)	–
Y soft 3D (Enterprise solutions, Czech Republic): Face shields (daily production capacity)	500
Weerg and PressUP (Italy): protective visors	500
BCN3D (3D Printing manufacturer): face shields	400
Formlabs (3D Printing company, United States): TEST swabs (potential capacity of 75,000–1,50,000 per day)	300

(continued)

Table 1 (continued)

Names of 3D printing company	No. of parts produced
Photocentric (Photo-polymer manufacturer, UK): face shield parts (first batch of prints; potential daily capacity for 4,860 parts)	200
Omni 3D (Industrial 3D Printing, Poland): face shields (daily capacity)	120

Table 2 PPE kit components and their importance

No.	Name of component	Importance
1	Gown	Gowns provide protection from the contamination of clothing with any infectious pathogens
2	Gloves	Gloves provide protection while handling pathogens or infected patients or contaminated surfaces
3	Mask	Surgical masks provide protection to avoid infectious pathogens enter the body through air. It works as a filter for the air before inhalation via nose or mouth
4	Goggles	Goggles provide protection to the eyes from direct contact with the pathogens
5	Headcover	Headcover acts as a barrier to isolate the headspace from the contaminated environment or surrounding
6	Face shield	A face shield is used to provide protection to facial skin and create a barrier and stop the direct infiltration
7	Shoe cover	Shoe covers provide protection to the feet so that it works as a barrier for the accumulation of pathogens or contamination of the human body

Some of the development in personal protective equipment are discussed in detail as under.

3.1 3D Printed Masks

Various 3D printing resource platforms, as well as many 3D printing companies, proposed multiple and a variety of models for a mask. This was done to facilitate other 3D printing companies of this sector to help by producing more number of masks, increasing the production. The Veterans Health Administration (VHA) designed a 3D printed surgical Stopgap mask for health care officials and workers. This mask can be reused after replacing some of the components which are readily available in the market. There is a filter membrane that requires disinfection by some solution soaking to serve well in this COVID-19 emergency (FDA Approves First 3D-Printed Mask for COVID-19 Support 2020; Tarfaoui et al. 2020) (Fig. 1).

The mask consists of three reusable 3D printed (3DP) parts namely the base of the mask, filter grill, and filter membrane as shown in Fig. 2. The mask base is supposed

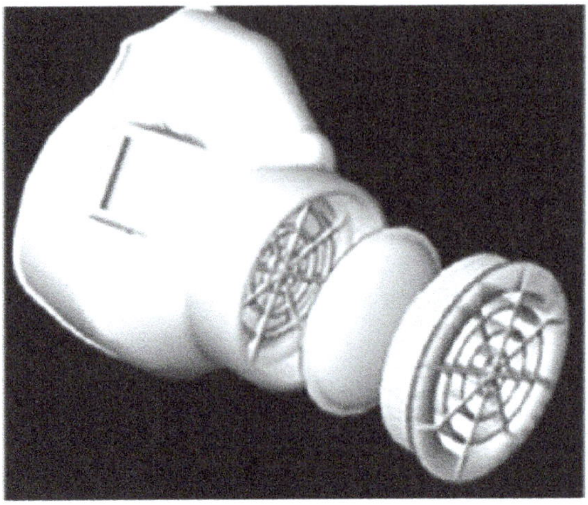

Fig. 1 The first prototype of a 3D printed mask during the pandemic (Tarfaoui et al. 2020)

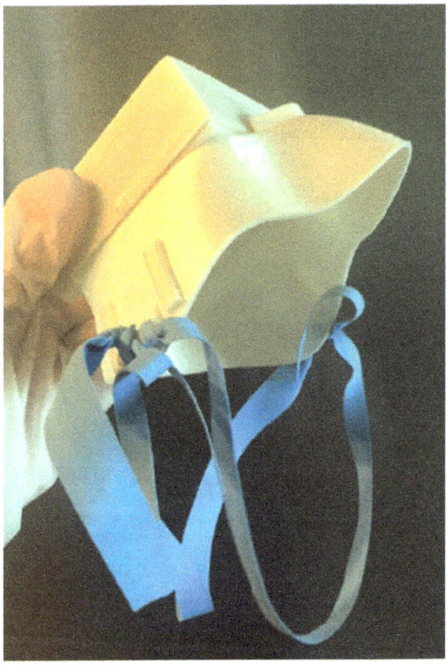

Fig. 2 3D printed masks are designed for level-I orthopedic trauma practice during COVID-19 situation (Claire et al. 2020)

to be in direct contact with the user's face. This mask needs to be soaked in slightly warm water to make it lightly malleable. This would help in getting it perfectly fit on the user's face when pressed gently. A snap mechanism is provided for the filter grill to fit in. The filter membrane holds up the filter paper against the filter grill and plays a crucial part in the entire 3DP mask making it reusable.

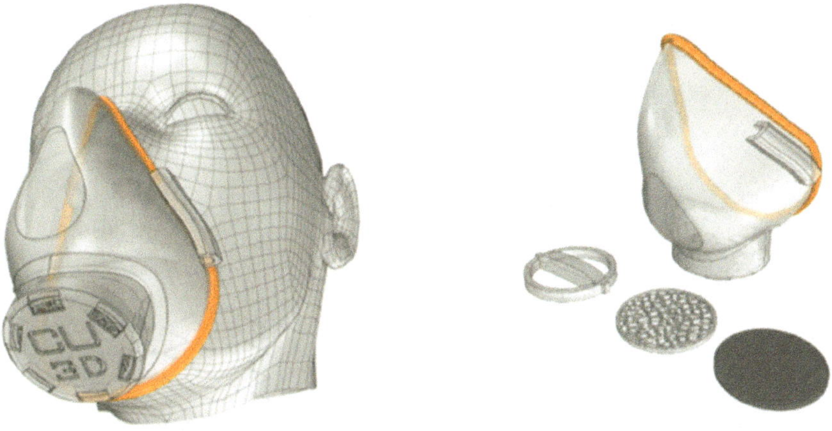

Fig. 3 NanoHack mask 3D printed with a recyclable as well as a biocompatible polymer, i.e., PLActive (Tarfaoui et al. 2020)

In recent days, Copper3D has shared a design mask viz NanoHack 2.0, as shown in Fig. 3 (Hack the Pandemic 2020). The mask consists of a strong and hermetic monoblock structure which is 3D printed with PLActive to avail maximum protection by isolating nose and mouth from the external contaminated environment. PLActive is a recyclable, as well as a biocompatible polymer that has copper nanocomposite, showing antimicrobial properties. There is a separate mechanism to make the mask watertight. So, the frame is sealed using a 3D printed edge with MDFlex, which again is an antimicrobial TPU. MDFlex is an innovative nanocomposite developed with a high-quality TPU98A and a patented nano-copper additive, scientifically proven to be effective (Tarfaoui et al. 2020).

The other very popular type of mask is the N95 respirators masks. These masks have been recommended by the Centers for Disease Control and Prevention (CDC) as they are more than 95% efficient at filtering 0.3 mm airborne particles (Rowan and Laffey 2020; Cai et al. 2018). Figure 4 shows the proposed design and the importance of each part of the mask. The design specifies that it can be 3D printed to make sure the sealing mechanism for the isolation of the nose and mouth from the external environment is taken care of. Mostly, fused deposition modeling (FDM), laser sintering (LS), and selective laser sintering (SLS) are used for the manufacturing of such masks.

Approaching the costliest solution using 3D printed products, full-faced snorkel masks have come into the picture (O.R. emergency only adaptations of full face masks to address COVID-19 CRISIS 2020; Robinson 2020). These masks were modified to take care of the norms laid by the CDC and WHO as shown in Fig. 5. Basically, these masks were implemented with the thought of minimizing the fogging problems and CO_2 rebreathing. For a longer period of usage, it has its own advantages and disadvantage. Some advantages include self-defogging, avoiding CO_2 rebreathing, strong shield for isolation, full coverage, as well as no direct contact with the equipment

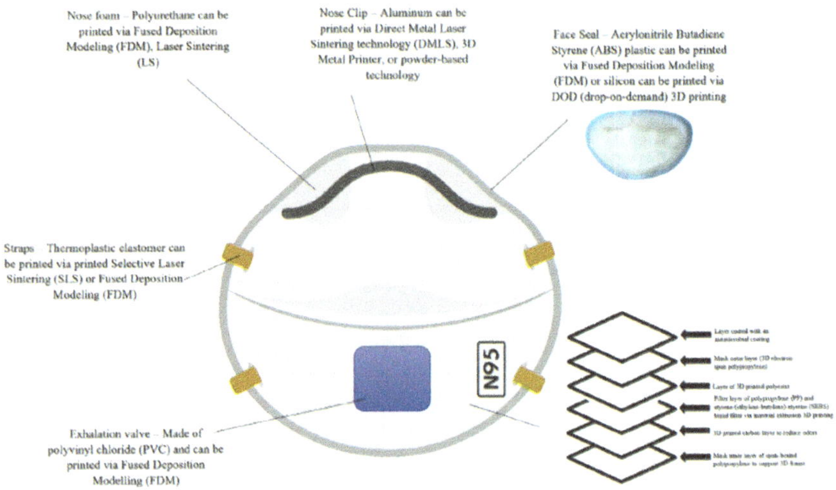

Fig. 4 A potential N95 3D printed mask prototype (Ishack and Lipner 2020)

Fig. 5 Snorkel mask with adapter and filters (Nicholson et al. 2020)

concerning nose and mouth (O.R. emergency only adaptations of full face masks to address COVID-19 CRISIS 2020; Robinson 2020; Nicholson et al. 2020). On the contrary, this mask is way too heavier for long-duration usage.

Fit testing is usually performed using the OSHA 29CFR1910.134 protocol in the PortaCount 8030 (Nicholson et al. 2020). The PortaCount clear tubing had a 5 mm outer diameter and a 3 mm inner diameter. The fit test includes various types of breathing such as normal breathing, deep breathing, breathing with different directional head moments, talking, grimace, breathing while bending over, etc., to test all sorts of favorable conditions. A passing fit factor for full-face masks like this design is 500, while for half-face masks is 100. The fit factor is calculated as the challenge aerosol concentration outside the respirator with respect to the challenge aerosol concentration that leaks inside the respirator during any fit test. Therefore, the higher the number it's better. Hence, with a little bit of trouble, this proves to be the costliest yet the safest option to opt for the high infection-prone individuals such as doctors and hospital staff. All that so far we have discussed about various face masks that can serve the purpose to provide safety to the users. But, once it is used, what after that? A new problem arises of disposal. When we think about the aspect of disposal, there are a few major points that should be kept in mind. Firstly, the disposal should be done properly as it can also be a medium to transmit the disease further. To do so, burning them or incineration can be done so that the pathogens are also dead. Still, proper care needs to be taken while burning so that it does not contaminate the surrounding. Secondly, such hazardous items can be disposed of in landfills. This would increase the cost factor drastically. But, the average amount of masks used per individual is 1 or more than 1 per day and in the case of hospital staff, the PPE kits they use are for single use only. So, imagining the amount of waste generated, reusage of masks will be better at least in the case of the general public.

Keeping the abovediscussed issues in mind, some scientists from different laboratories and organizations have tried finding a solution to manufacture reusable masks with the help of nanotechnology. Kolos Molnar et al. and his team from Hungary have proposed a type of mask with changeable filters as well as biodegradable (He et al.). These filters are nanoporous in nature. They are manufactured using the electrospinning technique (as shown in Fig. 6). The nanoporous structure of the filters allows only air molecules to pass through it. As these nano filters are made up of polylactic acid (PLA), they are transparent. The ideology behind this was to avoid

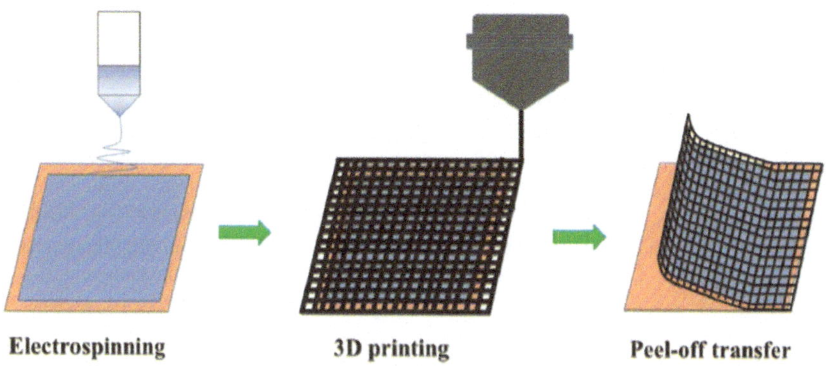

Fig. 6 The process steps involved in the manufacturing of nanoporous and biodegradable filters (https://doi.org/10.18063/ijb.v6i4.278)

the psychological trauma of covered faces in the surrounding at all times. This also allows people to exchange their expressions, basically implying to take care of the mental health of every individual.

3.2 3D Printed Face Shields

Face shields or face visors, as shown in Table 2, are used to isolate the overall face from the contaminated environment. To do so, 3D printing or additive manufacturing has helped a lot to accomplish the requirement to the fullest efficiency. Mostly the face visors are made up of comparatively cheaper material than any other PPE components as mentioned in Tables 2 and 3. Face visors have two main components namely, the head strap and the transparent cover. It needs a transparent layer to cover the face, as well as allow the user to view the external environment clearly. So, the materials used for the transparent shield is mostly polycarbonate or polyester film (as shown in Figs. 7 and 8).

Here, both the components can easily be manufactured using 3D printing as they don't require any special characteristics. The basic characteristics required in this case are flexibility, strength, rigidity, and lightweight. All the abovementioned properties can be easily obtained using the 3D printing technique, namely FDM (used here). It has been also observed that some engineers and people have tried manufacturing masks at their in-house 3D printers (de Araujo Gomes et al. 2020). This lowers down the possibility of contamination of the masks externally during transportation and other means.

As shown in the above Fig. 9a, ABS Premium MG94 filament was used as an input for the 3D printer. As per the given input as an STL file, the 3D printer takes around 4 h to completely print one component. Once the headband is ready (as shown

Table 3 Recently manufactured 3D printed in the COVID-19 pandemic

Product	Demand	Users	3D model source
Face mask	Very high	Frontline workers, doctors, hospital staff, the general public	https://amaskforall.com/
Face shield	Very high	Frontline workers, doctors, hospital staff, the general public	https://www.prusaprinters.org/prints/25857-prusa-face-shield https://www.3dsystems.com/covid-19-response#faceShield
Venturi valve	Medium	Hospitals	https://grabcad.com/library/respirator-free-reanimation-venturi-s-valve-1/details?folder_id=8017467
Nasal swab	Very high	Hospital staff, testing centers, and patients	https://usf.app.box.com/s/wxmlj0r66vp8bzei6o7sur1kq1jr8o1i/folder/109236323102

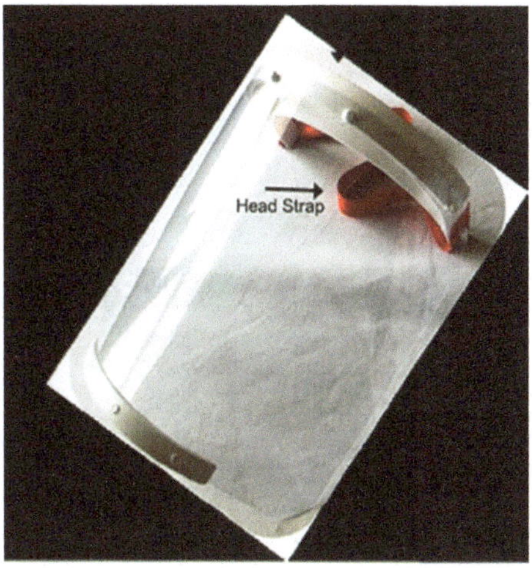

Fig. 7 A common design of reusable face shield or face visor (Armijo et al. 2020)

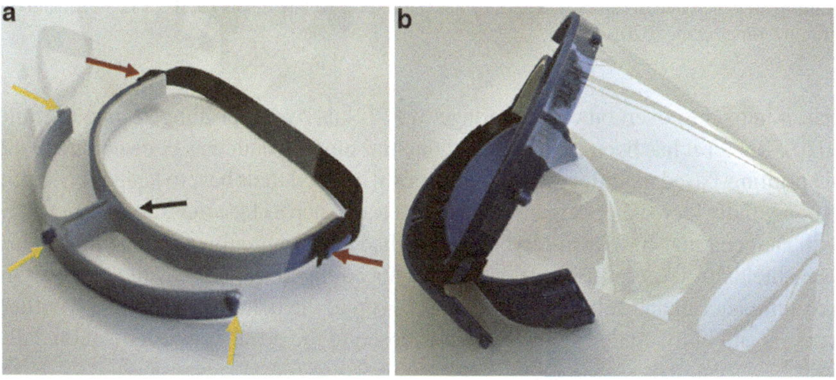

Fig. 8 **a** Inferior and **b** lateral views of an assembled face shield reported by Amin et al. (2020)

in Fig. 9b), the transparent film can be attached to the headband to obtain the final product. Hence, within 4 h of time, anyone can make a 3DP face shield at their own home (as shown in Fig. 9c).

4 Conclusion

The WHO had anticipated that PPE production would need to be ramped up by almost 40% to overcome the expected volumes required for patients, as well as COVID-19 warriors (World Health Organization 2020). Looking into the overall scenario, 3D printing has been very useful for some of the developments in PPE kits. Not only

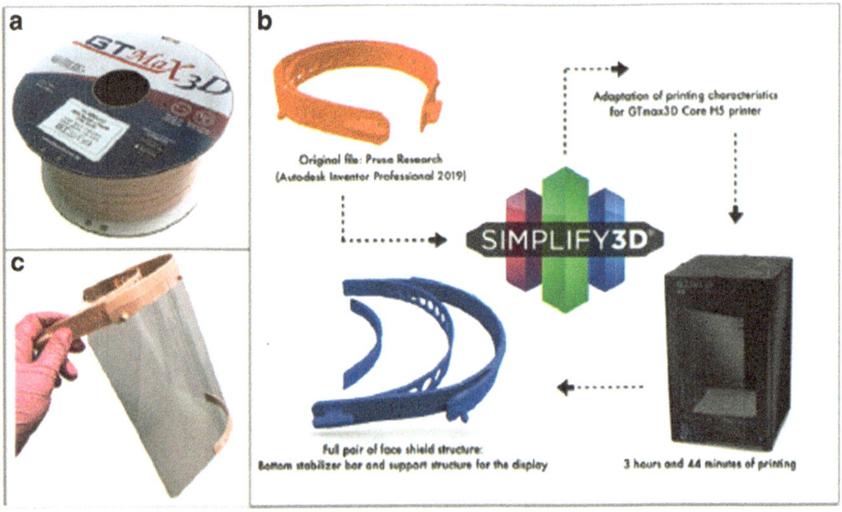

Fig. 9 **a** ABS premium MG94 filament by Gtmax3D (Americana, Sao Paulo, Brazil), **b** in-house printing flowchart and, **c** full face shield: Printed structure, acetate display and elastic headband (de Araujo Gomes et al. 2020)

that, it helped to overcome the shortage of PPE kits over a prolonged time (Armijo 2020). Rather it has been also observed that the supply chain was boosted due to the 3D printing-based manufacturing companies who tried their best to keep up with the exponentially increasing demand. In certain cases, such as face masks and face visors with nanocoatings have also served surplus than the commonly used N95 masks and face shields in terms of protection. The nanocoatings have also availed properties like antibacterial and self-cleaning. This adds up to the better protection of frontline workers and individuals who have been exposed to the deadly environment on daily basis.

In terms of effective manufacturing and efficient processing technique, FDM is mostly preferred for the production of face shields (Wesemann 2020). The major benefits of this process are that it has a very low post-processing requirement, as well as turns out to be the quickest among the other techniques. The next preferred technique is SLA, but it requires post-processing which includes removal of uncured resin, followed by cleaning and polishing the surface of the construct (Szykiedans and Credo 2016; Ćwikła et al. 2017).

The only issue that troubles the whole system is the cost-effectiveness. In certain cases like face visors, it is very cheap and affordable for everyone. But, some other masks such as the full-face snorkel mask (as shown in Fig. 5) are expensive. As a whole, 3D printing turned out to be effectively filling the gap in the supply chain, as well as enhancing the efficiency in providing protection to the frontline workers and other individuals.

References

Akmal JS, Salmi M, Mäkitie A, Björkstrand R, Partanen J (2018) Implementation of industrial additive manufacturing: intelligent implants and drug delivery systems. J Funct Biomater 9:41. https://doi.org/10.3390/jfb9030041

Amin D, Nguyen N, Roser SM, Abramowicz S (2020) 3D printing of face shields during COVID-19 Pandemic: a technical note. J Oral Maxillofac Surg 78(8):1275–1278. https://doi.org/10.1016/j.joms.2020.04.040

Armijo PR (2020) 3D printing of face shields to meet the immediate need for PPE in an anesthesiology department during the COVID-19 pandemic. Am J Infect Control 1–7:2020. https://doi.org/10.1016/j.ajic.2020.07.037

Armijo PR, Markin NW, Nguyen S, Ho DH, Horseman TS, Lisco SJ, Schiller AM (2020) 3D printing of face shields to meet the immediate need for PPE in an anesthesiology department during the COVID-19 pandemic. Am J Infect Control. https://doi.org/10.1016/j.ajic.2020.07.037

Cai M, Li H, Shen S (2018) Customized design and 3D printing of face seal for an N95 filtering facepiece respirator. J Occup Environ Hyg 15(3):226–234. https://doi.org/10.1080/15459624.2017.1411598

Claire N, Thomas BS, Lisa K, Schroder BS, Peter A, Cole MD (2020) Ten days to implementation of 3D-printed masks for a level-I orthopaedic trauma practice during the COVID-19 Pandemic. J Bone Joint Surg Am 95(1–6):102. https://doi.org/10.2106/jbjs.20.00881

Ćwikła G, Grabowik CW, Kalinowski K, Paprocka I, Ociepka P (2017) The influence of printing parameters on selected mechanical properties of FDM/FFF 3D-printed parts. IOP Conf Ser Mater Sci Eng 227:12033

de Araujo Gomes B, Queiroz FLC, Pereira PLO, Barbosa TV, Tramontana MB, Afonso FAC, Garcia EDS, Borba AM (2020) In-house three-dimensional printing workflow for face shield during COVID-19 Pandemic. J Craniofac Surg 31(6):652–653. https://doi.org/10.1097/scs.0000000000006723

Examplatory use of 3D printing to provide medical supplies during coronavirus (COVID-19) Pandemic in 2020. https://www.statista.com/statistics/1107198/covid-19-3d-printing-medical-supplies, Accessed 23 Nov 2020

FDA Approves First 3D-Printed Mask for COVID-19 Support. https://www.governmentciomedia.com/fda-approves-first-3d-printed-mask-COVID-19-support, Accessed 10 Nov 2020

Freeman S, Ramos R, AlexisChando P, Zhou L, Reeser K, Jin S, Soman P, Ye K (2019) A bioink blend for rotary 3D bioprinting tissue engineered small diameter vascular constructs. ActaBiomaterial 1–13. https://doi.org/10.1016/j.actbio.2019.06.052

Hack the Pandemic. https://copper3d.com/hackthepandemic/, Accessed 10 Nov 2020

He H, Gao M, Illés B, 3D printed and electrospun, transparent, hierarchical polylactic acid mask nanoporous filter. Int J Bioprint 6(4): 278. https://doi.org/10.18063/ijb.v6i4.278

Hoffman W (2006) Preparing for pandemic. Traffic World 270:10–12

Ishack S, Lipner SR (2020) Applications of 3D printing technology to address COVID-19—related supply shortages. Am J Med 133(7):771–773. https://doi.org/10.1016/j.amjmed.2020.04.002

Kumar S, Chandra C (2010) Supply chain disruption by avian flu pandemic for US companies: a case study. Transp J 49:61–73. https://doi.org/10.2307/40904915

Li X, Wang W, Zhao X, Zai J, Zhao Q, Li Y, Chaillon A (2020) Transmission dynamics and evolutionary history of 2019-nCoV. J Med Virol 92:501–511. https://doi.org/10.1002/jmv.25701

Matai I, Kaur G, Seyedsalehi A, McClinton A, Laurencin CT (2020) Progress in 3D bioprinting technology for tissue/organ regenerative engineering. Biomaterials 226. https://doi.org/10.1016/j.biomaterials.2019.119536

Nicholson K, Henke-Adams A, Henke DM, Kravitz AV, Gay HA (2020) Modified full-face Snorkel mask as COVID-19 personal protective equipment: quantitative results. Preprints, pp 1–15. https://www.preprints.org/manuscript/202004.0293/v1

O.R. emergency only adaptations of full face masks to address COVID-19 CRISIS (2020). https://oceanreefgroup.com/covid19/, Accessed 14 Apr 2020

Pettersson A, Salmi M, Vallittu P, Serlo W, Tuomi J, Mäkitie AA (2020) Main clinical use of additive manufacturing (three-dimensional printing) in Finland restricted to the head and neck area in 2016–2017. Scand J Surg 109:166–173. https://doi.org/10.1177/1457496919840958

Robinson F (2020) Self-protection: how NHS doctors are sourcing their own PPE. BMJ 369(May):8–11. https://doi.org/10.1136/bmj.m1834

Robotic Surgery: Custom 3D Printed Surgical Instruments for Every Patient (2020). https://blog.peekmed.com/robotic-surgery-3d-printed-instrument/, Accessed 20 Nov 2020

Rowan NJ, Laffey JG (2020) Science of the total environment challenges and solutions for addressing critical shortage of supply chain for personal and protective equipment (PPE) arising from coronavirus disease (COVID19) pandemic—case study from the Republic of Ireland. Sci Total Environ 725:138532. https://doi.org/10.1016/j.scitotenv.2020.138532

Salmi M, Akmal JS, Pei E, Wolff J, Jaribion A, Khajavi SH (2020) 3D printing in COVID-19: productivity estimation of the most promising open source solutions in emergency situations. Appl Sci (Switzerland) 10(11):1–15. https://doi.org/10.3390/app10114004

Szykiedans K, Credo W (2016) Mechanical properties of FDM and SLA low-cost 3-D prints. Procedia Eng 136:257–262. https://doi.org/10.1016/j.proeng.2016.01.207

Tarfaoui M, Nachtane M, Goda I, Qureshi Y, Benyahia H (2020) 3D printing to support the shortage in personal protective equipment caused by COVID-19 Pandemic. Mater (Basel, Switzerland) 13(15):3339. https://doi.org/10.3390/ma13153339

Wang Z, Kumar H, Tian Z, Jin X, Holzman JF, Menard F (2008) Visible light photoinitiation of cell-adhesive gelatin methacryloyl hydrogels for stereolithography 3D bioprinting. ACS Appl Mater Interfaces 10(32):26859–26869. https://doi.org/10.1021/acsami.8b06607

Wesemann C (2020) 3-D printed protective equipment during COVID-19 pandemic. Mater (Basel) 13(8):1–9. https://doi.org/10.3390/MA13081997

World Health Organization (2020) WHO Coronavirus Disease (COVID-19) Dashboard. https://covid19.who.int/table, Accessed 23 Nov 2020

World Health Organization (2020) Shortage of personal protective equipment endangering health workers worldwide. https://www.who.int/news-room/detail/03-03-2020-shortage-of-personal-protectiveequipment-endangering-healthworkers-worldwide, Accessed 1 Nov 2020

Zhang X, Mao J, Zhou Y, Ji F, Chen X (2020) Mechanical properties and osteoblast proliferation of complex porous dental implants filled with magnesium alloy based on 3D printing. J Biomater Appl. https://doi.org/10.1177/0885328220957902

Use of Additive Manufacturing in the Battle Against COVID-19

Harish Kumar Banga, Rajesh Kumar, Parveen Kalra, R. M. Belokar, and Somya Saxena

Abstract As the coronavirus pandemic has spread across the world, travel restrictions, social distancing measures, and working from home are in place in every country. Health services even in developed nations are facing unprecedented demand not only for hospital beds and ventilators but also for clinical staff and the personal protective equipment (PPE) they require to treat patients. Shortages of equipment have led to greatly increased activity by the global network of 3D printing enthusiasts, offering their designs for free in the hope of reducing infections. In this chapter, few examples of 3D-printed items include face shields, nasal swabs, respirator masks and ventilator components used by folk's has been discussed.

Keywords 3D printing · Additive manufacturing · COVID-19 · Human life · Innovation

1 Introduction

With 3D printing it has become feasible to rapidly design, prototype and produce an idea, not in days or weeks, but sometimes in hours. As we navigate the COVID-19 pandemic, the following examples show how 3D printing has the potential to solve supply chain problems caused by the overwhelming demand for clinical equipment (https://copper3d.com/hackthepandemic/).

With the need to ramp up coronavirus testing globally, manufacturers of the long thin nasal swabs made from artificial materials have been struggling to keep up with the demand (Banerjee et al. 2019).

H. K. Banga (✉)
Department of Mechanical Engineering, GNDEC, Ludhiana, India

R. Kumar
Department of Mechanical Engineering, UIET, PU, Chandigarh, India

P. Kalra · R. M. Belokar
Department of Production and Industrial Engineering, PEC, Chandigarh, India

S. Saxena
Department of Physical Medicine & Rehabilitation, PGIMER, Chandigarh, India

K. Sandhu et al. (eds.), *Emerging Applications of 3D Printing During CoVID 19 Pandemic*, Lecture Notes in Bioengineering, https://doi.org/10.1007/978-981-33-6703-6_9

HP has developed a 3D-printed hands-free door opener and a mask adjustor clasp to improve performance of masks worn for long periods. Hospital-grade FFP3 face masks will soon be at the production stage, and 3D-printed ventilator components are in development. HP and its partner companies are making designs for simple components available for free (Bachtiar et al. 2020). 3D printing practitioners who wish to be a part of the collaborative effort against COVID-19 can contribute new ideas via HP's dedicated 3D printing website. Dassault Systèmes is using scientific simulation of the human sneeze to design and develop PPE via their 3DEXPERIENCE Lab OPEN COVID-19 online community (Banga et al. 2020a). Dassault's simulator functions, already used in the aerospace and automobile industries to generate a dynamic simulation of fluid and air flow, are now being used to understand the physics of sneezes (Banga et al. 2020b).

2 3D Printing Resources and Covid 19 Projects

2.1 Ventilator Valves

In March 2020, a shortage of valves for ventilators for 250 Covid patients at a hospital in Brescia, Italy led to a collaboration between Issinova and Lonati, who were able to rapidly print the valves designed by Issinova's Cristian Fracassi (Fig. 1) (https://3dprintingindustry.com/news/3d-printing-community-responds-to-covid-19-and-coronavirus-resources-169143/).

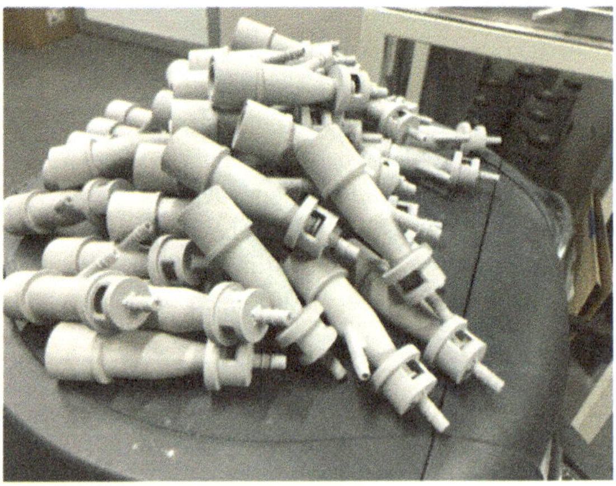

Fig. 1 3D Printed Oxygen Valves (https://3dprintingindustry.com/news/3d-printing-community-responds-to-covid-19-and-coronavirus-resources-169143/)

Fig. 2 Additive manufactured hands-free door handle (https://3dprintingindustry.com/news/3d-printing-community-responds-to-covid-19-and-coronavirus-resources-169143/)

2.2 3D-Printed Hands-Free Door Handle Attachment

The danger of COVID-19 transmission through contact with surfaces in the workplace led to development of a hands-free door handle attachment (Banga et al. 2017a, 2018, 2020c). As shown in Fig. 2, the attachment enables the door to be opened using the elbow and forearm. The design files are available to download for free.

2.3 Additive Manufactured Hand Sanitizer Holder

A 3D-printable holder for a hand sanitizer container to be worn on the wrist is shown in Fig. 3. The simple design was realized by Moath Abuaysha, He aims to cleanse hands globally of the Coronavirus. He specializing in surgical 3D printing in Saudi Arabia has designed a 3D printable wrist clasp to hold a sanitizer bottle for easy access Banga et al. (2014, 2017a, b, 2018).

2.4 3D Printing Service

Protolabs is an additive manufacturing company offering a combined design optimisation, 3D printing and CNC machining service. The business is well known in the additive manufacturing world and its capabilities are being used in the current

Fig. 3 3D printed hand sanitizer clasp (https://3dprintingindustry.com/news/3d-printing-community-responds-to-covid-19-and-coronavirus-resources-169143/)

COVID-19 pandemic. The organisation's Twitter account provides examples of objects produced, illustrating how virtual creation can provide rapid response in a crisis. Recently some of their clients have told us of their assistance in producing COVID-19 packs and ventilators (Kumar et al. 2020; Cai et al. 2018; https://www.who.int/emergencies/diseases/novel-coronavirus-2019).

2.5 3D Print Farms

Barcelona-based BCN3D has used 63 3D printing machines to supply clinical instruments. California-based Airwolf 3D apply their own remarkable array of 3D printers. The business similarly provides remote guidance for the clinical workforce to see the potential of 3D printing as shown in Fig. 4. More than 5,000 pairs of 3D-printed accredited goggles have been given to pandemic-affected regions to date, with the partnership making a further 2,000 currently. The goal is to increase each day to 10,000 sets in the coming days and weeks (https://lowellmakes.com/covid-19-response/; D'Urso et al. 1998; Tino et al. 2020; https://www.thingiverse.com/thing:4177128).

2.6 3D-Printed Quarantine Hospital Booths

Winsun, a Chinese company that specialises in the 3D printing of buildings, has dispatched 15 3D-printed quarantine booths (Fig. 5) to the Xianning Focal Emergency facility in the Hubei Region. Manufacture had to be outside Wuhan, to avoid

Fig. 4 Flashforge guider II print farm (D'Urso et al. 1998)

Fig. 5 3D printed quarantine booths (Guideline for Disinfection and Sterilization in Healthcare Facilities 2008)

the spread of COVID-19. The shortage of hospital beds became a real issue for clinical staff as the number of patients grew exponentially. The inside parts of the booths are decorated and have their own water and power supply (Guideline for Disinfection and Sterilization in Healthcare Facilities 2008).

2.7 *Additive Manufactured Face Visors*

The World Health Organisation provides excellent guidance concerning facemasks. In use a facemask accumulates germs, and moist components can harbour them if they are not adequately cleaned or prepared. The visor shown in Fig. 6 has a 3D-printed frame and clear plastic shield to protect the wearer. Some are provided free of charge (https://www.agorize.com/en/challenges/code-life-challenge?lang=en).

Fig. 6 Stratasys additive manufactured face shield frame (https://www.thingiverse.com/thing:4222563)

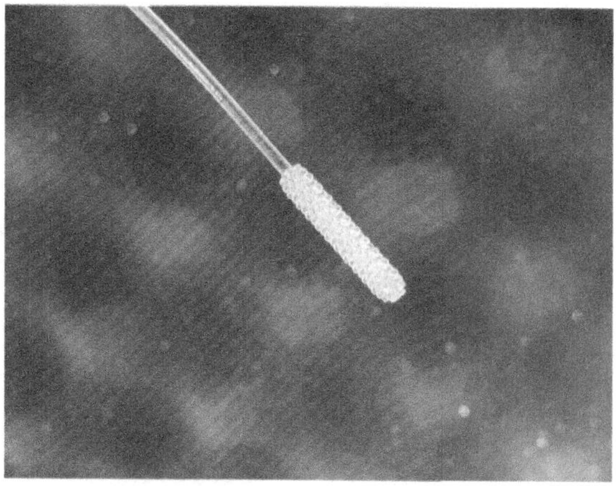

Fig. 7 Close-up of the 3D printed testing Swab (https://www.materialise.com/en/hands-freedoor-opener/technical-information)

2.8 Additive Manufactured COVID-19 Test Swabs for Hospitals

Formlabs, a Boston-based 3D printing company, is endeavouring to produce 3D-printed COVID-19 test swabs, It is employing 1,000 printers to mass-produce swabs. More than 300 test swabs can be produced in one 11-h print cycle, allowing Formlabs to make 75,150 swabs to date (https://www.materialise.com/en/hands-freedoor-opener/technical-information; https://www.3dprintingmedia.network/covid-19-3dprinted-valve-for-reanimation-device/; https://doi.org/10.1097/dss.0000000000002378). Formlabs is working with 3 leading U.S. clinical workplaces and educator Dr. Ramy Arnaout on the swab method (Fig. 7).

2.9 3D-Printed Components to Automate Pumping for Retrofitting Manual Ventilators

The Mechanical and Advanced Aircraft Structure Division at UC San Diego have developed 3D-printed parts to automate manual ventilators. These control the speed and extent of the compressions used to assist patient breathing, as shown in Fig. 8.

2.10 Stop-Gap Additive Manufactured mMask

3D printing of face mask components on FDM/FFF printers using PLA and other feedstocks has been achieved by designing a structure suitable for 3D printing. SLS and biocompatible nylon has been used to 3D print a stop-gap facemask (SFM), held in place with elastic. Modelers and clinicians

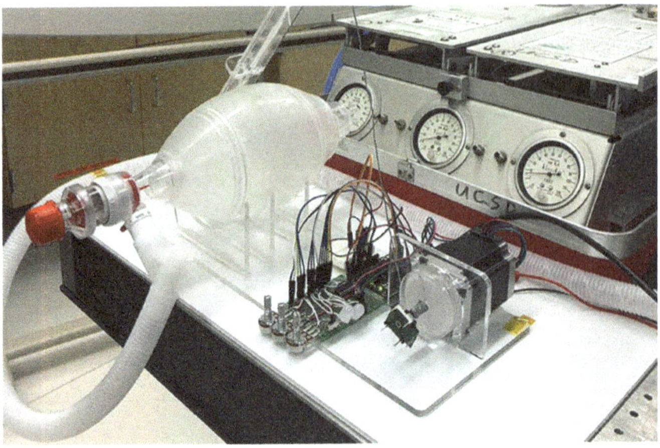

Fig. 8 Using 3D printed parts and off-the-shelf components (https://3dprintingindustry.com/news/3d-printing-community-responds-to-covid-19-and-coronavirus-resources-169143/)

from the US Veterans Wellbeing Organization have used 3D-Frameworks to develop a 3D cover to fit the face. The SFM is a printed cover that holds in place a square of filter material (Ramanathan et al. 2020; https://grabcad.com/library/respirator-free-reanimation-venturi-s-valve-1; https://doi.org/10.1056/nejmp2006141; https://no2covid.com; https://multimedia.3m.com/mws/media/9885560/tdb-cleaning-reusable-respirators-and-papr-assemblies.pdf; https://www.3dprintingmedia.network/personalizedppe-mask/). The SFM may be cleaned using disinfectants or autoclaved. The SFM will be used for clinical isolation zones when standard PPE is unavailable or for lesser non-clinical conditions. Evaluation of the SFM (Fig. 9) in a clinical setting is planned.

3 Discussion

3D-printing has been rapidly applied to help in the supply of PPE and other medical equipment for the COVID-19 pandemic. It is essential that FFF/FDM 3D-printed items do not act as vectors to spread the virus. Intense UV light may be used as a disinfectant but it is not clear how long the COVID-19 virus persists on printed items. An article in the *BMJ* (*British Medical Journal*) notes that, of people infected with COVID-19, 50–75% have been asymptomatic. 3D printing has already been used to increase the supply of ventilators and could be applied more widely to produce large numbers in 'print farms'. Social media have helped individuals to work together in these endeavours.

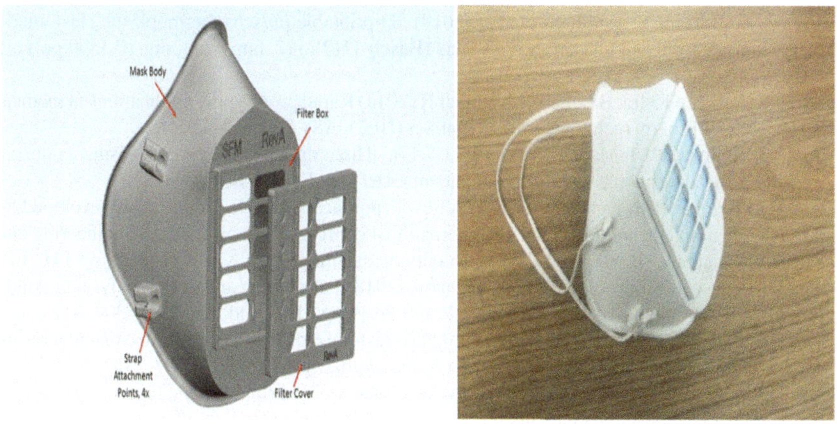

Fig. 9 The 3D systems stopgap face mask (https://3dprintingindustry.com/news/3d-printing-community-responds-to-covid-19-and-coronavirus-resources-169143/)

4 Conclusion

The magic of 3D printing consists in its potential both to develop replacement parts for machines and to rapidly create prototypes from scratch. It can massively affect the supply chain, filling the temporary shortfalls currently being experienced by almost all companies. In the current global pandemic, this includes methods for rapid manufacture of face shields, respirator masks, and ventilator parts. Design, prototype and production can take hours rather than days, weeks or longer, something that producers in general are not able to achieve. A number of larger manufacturers may be using plant-based materials as a 3D printing resource.

Printers have specific requirements in terms of temperature and are able to operate with a range of voltages. Although there is a steep learning curve to operating this system, many corporations are in a position to step up to this undertaking. It is important to maintain sterility of products, taking into account the place of manufacture and the packaging materials. Clinical oversight and proper certification is therefore required.

In conclusion, 3D printing is a necessary link in the chain between vendors and clients. This innovative method, with its ability to meet the present market fluctuations, is contributing to saving lives.

References

About NanoHack. https://copper3d.com/hackthepandemic/. Accessed 24 Mar 2020

Bachtiar EO, Erol O, Millrod M et al (2020) 3D printing and characterization of a soft and biostable elastomer with high flexibility and strength for biomedical applications. J Mech Beh of Biomed Materials 104: https://doi.org/10.1016/j.jmbbm.2020.103649

Banerjee SS, Burbine S, Kodihalli N et al (2019) 3D-printable pp/sebs thermoplastic elastomeric blends: preparation and properties. Polymers (Basel) 11(2):347. https://doi.org/10.3390/polym11020347

Banga HK, Parveen Kalra, Belokar RM, Kumar R (2014) Rapid prototyping applications in medical sciences. Int J Emerg Technol Comput Appl Sci (IJETCAS) 5(8):416–420

Banga HK, Belokar RM, Madan R, Dhole S (2017a) Three dimensional Gait assessments during walking of healthy people and drop foot patients. Def Life Sci J 2:14–20

Banga HK, Belokar RM, Kumar R (2017b) A novel approach for ankle foot orthosis developed by three dimensional technologies. In: 3rd international conference on mechanical engineering and automation science (ICMEAS 2017), University of Birmingham, UK, vol 8, no 10, pp 141–145

Banga HK, Parveen Kalra, Belokar RM, Kumar R (2018) Fabrication and stress analysis of Ankle foot orthosis with additive manufacturing. Rapid Prototyp J Emerald Publ 24(2):300–312

Banga HK, Kalra P, Belokar RM, Kumar R (2020a) Customized design and additive manufacturing of kids' ankle foot orthosis, Rapid Prototyp J. https://doi.org/10.1108/RPJ-07-2019-0194

Banga HK, Kalra P, Belokar RM, Kumar R (2020b) Effect of 3D-printed Ankle foot orthosis during walking of foot deformities patients. In: Kumar H, Jain P (eds) Recent advances in mechanical engineering, Lecture notes in mechanical engineering. Springer, Singapore, pp 275–288

Banga HK, Kalra P, Belokar RM, Kumar R (2020c) Improvement of human gait in foot deformities patients by 3D printed Ankle-foot orthosis. In: Singh S, Prakash C, Singh R (eds) 3D printing in biomedical engineering. Materials horizons: from nature to nanomaterials. Springer, Singapore

Cai M, Li H, Shen S et al (2018) Customized design and 3D printing of face seal for an N95 filtering facepiece respirator. J Occup Environ Hyg 15(3):226–234. https://doi.org/10.1080/15459624.2017.1411598

Code Life Ventilator Challenge. https://www.agorize.com/en/challenges/code-life-challenge?lang=en. Accessed 24 Mar 2020

Coronavirus disease (COVID-19) Pandemic. https://www.who.int/emergencies/diseases/novel-coronavirus-2019. Accessed 24 Mar 2020

COVID-19 Response. https://lowellmakes.com/covid-19-response/. Accessed 24 Mar 2020

D'Urso PS, Atkinson RL, Lanigan MW, Earwaker WJ, Bruce IJ, Holmes A, Barker TM, Effeney DJ, Thompson RG (1998) Stereolithographic (SL) biomodelling in craniofacial surgery. Br J Plast Surg 51:522–530

Guideline for Disinfection and Sterilization in Healthcare Facilities (2008). https://www.cdc.gov/infectioncontrol/pdf/guidelines/disinfection-guidelines-H.pdf

Hands-Free Door Openers. https://www.materialise.com/en/hands-freedoor-opener/technical-information. Accessed 24 Mar 2020

HEPA Mask. https://www.thingiverse.com/thing:4222563

https://3dprintingindustry.com/news/3d-printing-community-responds-to-covid-19-and-coronavirus-resources-169143/

Ishack S, Lipner SR. A review of 3-dimensional skin bioprinting techniques: applications, approaches, and trends [e-pub ahead of print]. Dermatol Surg. https://doi.org/10.1097/dss.0000000000002378. Accessed 25 Mar 2020

Italian hospital saves Covid-19 patients' lives by 3D printing valves for reanimation devices. https://www.3dprintingmedia.network/covid-19-3dprinted-valve-for-reanimation-device/. Accessed 24 Mar 2020

Kumar P, Sharma P, Banga HK, Kalra P, Belokar RM, Kumar R(2020) Performance evaluation of Ankle foot orthosis on lower extremity disabled persons while walking using OpenSim. In: Sachdeva A, Kumar P, Yadav P, Garg R, Gupta A (eds) Lecture notes on multidisciplinary industrial engineering, Springer, Singapore pp 192–210

Makers Guide. https://creality.com/info/makers-guide-3d-printed-face-maskno-worries-on-mask-shortage-and-virus-infection-i00248i1.html. Accessed 24 Mar 2020

Prusa Protective Face Shield—RC2. https://www.prusaprinters.org/prints/25857-protective-face-shield-rc1. Accessed 24 Mar 2020

Ramanathan K, Antognini D, Combes A, Paden M, Zakhary B, Ogino M, MacLaren G, Brodie D, Shekar K (2020) Planning and provision of ECMO services for severe ARDS during the COVID-19 pandemic and other outbreaks of emerging infectious diseases. Lancet Respir Med S2213–S2600(20):30121

Ranney ML, Griffeth V, Jha AK. Critical supply shortages—The need for ventilators and personal protective equipment during the covid-19 pandemic [e-pub head of print]. N Engl J Med. https://doi.org/10.1056/nejmp2006141. Accessed 25 Mar 2020

Respirator-free reanimation Venturi's valve (rev. 4). https://grabcad.com/library/respirator-free-reanimation-venturi-s-valve-1. Accessed 24 Mar 2020

The no2covid-ONE. https://no2covid.com. Accessed 2 Apr 2020

Technical Data Bulletin. Cleaning reusable respirators and powered air purifying respirator assemblies. https://multimedia.3m.com/mws/media/988556O/tdb-cleaning-reusable-respirators-and-papr-assemblies.pdf. Accessed 24 Mar 2020

Tino et al (2020) Easy COVID-19. 3D Print Med 6(11):7–8. https://www.isinnova.it/easy-covid19-eng/. Accessed 24 Mar 2020

Tino R, Moore R, Antoline S (2020) Flexible Mask Valve. 3D Printing in Medicine 6:11 https://doi.org/10.1186/s41205-020-00064-7. https://www.thingiverse.com/thing:4177128. Accessed 24 March 2020

WASP shares open source processes for production of personalized PPE masks and helmets. https://www.3dprintingmedia.network/personalizedppe-mask/

3D Printing During COVID-19: Challenges and Possible Solutions

Jyotindra Narayan, Sanchit Jhunjhunwala, and Santosha K. Dwivedy

Abstract The COVID-19 pandemic has led to the imminent collapse of medical supply chains across global economies at an unprecedented scale. Essential supplies such as personnel protective equipment (PPE), ventilator components, and face shields have witnessed a continuous rise in demand and eventually boost the role of 3D printing. Over the months following the spread of the pandemic, 3D printed alternatives of many medical devices were made more accessible to hospitals, mostly by community-sourced design and fabrication. However, with the high volume usage of additive methodologies, several challenges associated with the design, manufacturing, and deployment of medical products have now been brought to further attention. In this work, a systematic evaluation of such challenges along with few possible solutions are presented. The pandemic and its effects on the industry are introduced in the context of disruptions caused across the supply chains. The role of additive manufacturing to counter these effects is presented with an introduction of the technology itself. Employing 3D printed products to address the shortages of healthcare equipment are mentioned and visualized. Thereafter, a central discussion is followed on the issues arising from the shift in production methodology of such medical devices—from conventional manufacturing to additive one. The problems are highlighted by discussing two important types of COVID-19 related 3D printed medical equipments—mechanical ventilators and PPEs. Thereafter, few possible solution methodologies are discussed as case studies of two particular instances of having such problems. Finally, a conclusion is drawn to solve the issues raised using similar solutions followed by future opportunities.

Keywords COVID-19 · Personnel protective equipment · Mechanical ventilators · 3D printing

J. Narayan (✉) · S. Jhunjhunwala · S. K. Dwivedy
Department of Mechanical Engineering, Indian Institute of Technology Guwahati, Guwahati 781039, Assam, India

K. Sandhu et al. (eds.), *Emerging Applications of 3D Printing During CoVID 19 Pandemic*, Lecture Notes in Bioengineering, https://doi.org/10.1007/978-981-33-6703-6_10

1 Introduction

The COVID-19 pandemic has already evolved into an event with plausible anthropological consequences (Higgins et al. 2020). In an era otherwise noted for the unstoppable growth pace, this virus's outbreak has witnessed an unprecedented dampening of commotion. The effects are most seriously made tangible by overloading healthcare systems (Newsroom 2020) and the disruption of supply chains (McKibbin and Fernando 2020). Even though it is not the first of its kind, either in terms of how contagious it is (MERS, SARS) or in terms of how deadly it is (Ebola, MERS, Avian Flu) (Callaway et al. 2020), the SARS-CoV2 virus has been able to spread to more than 230 of the 251 countries and territories of the world, as recognized by the United Nations (WHO coronavirus disease (COVID-19) dashboard 2020). This is primarily because of the augmented connections and urbanized domains of human residence being dependent on each other. An implication of such dependency among the connections ultimately led to the chain reactions of repercussions done by the pandemic (Ivanov and Dolgui 2020). Therefore, it becomes crucial to rapidly address the issues faced by regions without requiring physical contact with other areas. This calls for decentralized operations of the economy with locally sufficient ecosystems for services, trade, and specifically, manufacturing (Mollenkopf et al. 2020).

Additive manufacturing (AM) is a category of manufacturing that uses various processes to combine elementary forms of material, such as layers, blocks, segments, or lines, to produce the final component (Gibson et al. 2014). The colloquial term "3D printing" is derived from the powder bed process developed to use cornstarch as a substrate material by MIT in 1993 (MIT News Office 2011). However, the term "3D printing" has grown to be associated with polymer-based AM processes mostly and, therefore, acceptable as a synonym for additive manufacturing in general. Materials used in the additive manufacturing of biomedical devices can be classified, depending upon the objective, in the form of feedstock (powder, liquid, and wire), biocompatibility (intended usage environment), and nature (metallic, plastics, and organic). The application of biocompatible materials such as Poly-ethylene glycol diacrylate, Gelatin methacrylate, MG63 cells, Alginate, and PCL electrospun scaffold allows for 3D bioprinting of medical products. AM allows for rapid prototyping of designs, with automated fabrication processes that convert the generated CAD model to required files such as STL and G-codes. This reduces the cost and time of development by eliminating manual labor needed for prototype making processes such as tooling (Kenney 2013). Beyond development, 3D printing can be significantly advantageous over other scaled production techniques when utilized in the right conditions (Berman 2012).

These benefits of 3D printing have enabled a significant need for the design, development, and manufacturing of biomedical products, such as implants, stents, drug tablets and patches, prosthetics, etc. (Liaw and Guvendiren 2017). It allows for improved development cycles in terms of both speed and flexibility in design. It also offers customizability, allowing patient-specific solutions at no changes in cost (Ventola 2014). This has caused an explosion in the usage of additive manufacturing technologies to create medical products, with such processes becoming commonplace and even fundamental to the production of many devices. For example, few orthopedic implants now rely upon selective laser melting (SLM) based production on ascertaining their standards for osteocompatibility (Nikolova and Chavali 2019). However, the inter-reliant bonding of material used and printing technique followed to fabricate the biomedical products limits the functional properties such as surface characteristics, strength, and biocompatibility. For instance, powder-based materials can only be printed using spray or jet based techniques.

As the healthcare industry has also been proactive in adopting 3D printing, the COVID-19 pandemic has also been addressed with a barrage of additively manufactured components and devices (Choong et al. 2020). For instance, 3D printed equipment for COVID-19 care ranges from ventilator components such as valves and adapters to entire ventilator assemblies (Tino et al. 2020). 3D printing based production has proved extremely beneficial for regions with the sudden scarcity of resources caused by widespread lockdowns - from personnel protective equipment such as face shields and respirators to masks for public usage for the resumption of socioeconomic activities (Manero et al. 2020). The CAD model of few such 3D printed products is shown in Fig. 1. This may be attributed to the relative ease of development and collaboration along with the independence of the process from manual labor and fixed supply chains.

In this chapter, a number of such applications of 3D printing have been investigated, focusing on the problems faced in their usage, development, or production.

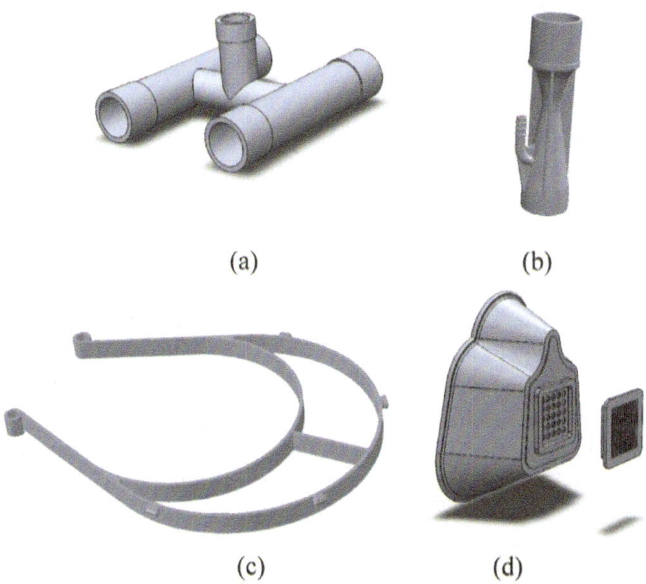

Fig. 1 Various 3D printed medical devices developed during COVID-19 pandemic. **a** Ventilator H connector, **b** ventilator venturi valve, **c** face shield headgear, and **d** mask with a replaceable filter

This is done with a thorough review of the literature published on various repositories about implementing 3D printing practices for producing essential medical supplies. Two specific types of products have been discussed to detail the problems faced with their manufacturing and establish a clear context for the associated issues. These products are related to the mechanical ventilation and personnel protective equipment kits used during COVID-19. Possible solutions for the listed problems are discussed in a detailed manner. The presented solutions are not sharing a direct relation with the mentioned problems; however, discuss the methodologies to draw inspiration while addressing the issues. Thereafter, subsequent opportunities for 3D printing in the emergence of health care products are briefly communicated. Finally, the conclusion is drawn from these case studies to provide a broader perspective of the solutions for the respective problems.

The following sections are organized in the following manner. Section 2 discusses the challenges reported during the additive manufacturing of medical products during the COVID-19 pandemic. In Sect. 3, two possible solutions to encounter various challenges are duly reviewed. Section 4 presents the future opportunities of 3D printing in the healthcare industry. At last, the conclusion of the current work is reported in Sect. 5.

2 Challenges Encountered 3D Printing During COVID-19

The shortage of essential medical supplies during the COVID-19 pandemic has prompted 3D printed alternatives to emerge for a diverse set of products. These products include PPE components such as face shield headgear and protective goggles, ventilator parts and assemblies, mask and respirators, and even consumer products such as door handle extensions and sanitizer dispensers (Attaran 2020). While many AM process types have been used to fabricate these products, the most commonly deployed process is fused filament fabrication (FFF). In contrast, methods like selective laser melting (SLM), laser polymerization (SLA) with digital light processing (DLP), and multi-jet fusion (MJF) have also been used widely (Tino et al. 2020). Cutting edge technologies such as CLIP (Continuous liquid interface production) have also been used in some instances (Novak and Loy 2020). This has implied widespread deployment of these relatively adolescent technologies into new product spheres. Therefore, some challenges have been encountered in the development of such products with these technologies. The issues range from technical problems that affect product quality, compatibility, and reliability; to economic challenges such as scalability; to legal and ethical issues associated with the usage of the developed products, such as regulatory approvals, intellectual property, and certification. To explore these contextually, the following cases have been studied.

2.1 3D Printed Ventilators: Valves, Adapters, and Splitters

There is a widespread demand for ventilators and their components because of the nature of the disease. The SARS-CoV2 causes difficulty breathing as it infects and damages epithelial cells within the lung (Moein et al. 2020). This reduces the gaseous exchange rate from the inhaled air, which causes oxygen levels to drop in the patient's body. With further and prolonged damage, it becomes difficult for the person to inhale and exhale, requiring breathing assistance (Goyal et al. 2020). The resulting demand for ventilation systems and their components across medical markets, coupled with stifled supply chains, causes widespread shortages (Truog et al. 2020). 3D Printing allows local sourcing of required components or entire systems themselves, overcoming a broken supply chain. For example, valves, tubing connectors, adapters, etc., are a few ventilator components widely replaced with 3D printed alternatives when necessary (Ishack and Lipner 2020).

However, there are a few concerns with using this fabrication methodology to produce such critical care components. One of the significant limitations in adopting 3D printing to any manufacturing pipeline is the narrow range of materials used to develop a ventilator component. As a result, these limitations are driven by the process of additive manufacturing selected for the task and the produced components' usage conditions. As a result, especially in intrusive medical devices, the availability of material choice is greatly restricted to a few classes of materials, which are sometimes inaccessible or unaffordable (Shahrubudin et al. 2020). Secondly, medical ventilators are heavily regulated products that require certification for the slightest of alterations made to either the design or the fabrication process. Apart from an emergency implementation, the acceptance of 3D printed alternatives to otherwise approved devices is challenging without appropriate testing and validation of each produced component. With the advantage of 3D printing relies on the speed at which such alterations and changes may be made, the time required for testing and regulatory approvals diminishes this benefit. This renders the limited scalability of existing 3D printing methods for developing medical components. Although the advent of 3D printing offers widespread community collaboration in research, as well as setups several open source projects to propagate the free flow of product design features, this also guarantees an imminent clash with intellectual property rights of developers from different domains having similar innovative ideas.

Moreover, conventional manufacturing methods are better suited for mass production of components, given their characteristics of repeatability and reduction in cost per component, increasing the number of components produced as visualized in Fig. 2 (Rengier et al. 2010; Busachi et al. 2017). Therefore, once supply chains are rebuilt after the pandemic, 3D printing shall no longer remain viable as a means of production en masse for valves, adapters, and splitters. Further, sizes of components or batches of components produced by 3D printing techniques depend upon the workspace area of the fabricating machine, which are mostly limited to small areas in the range of a few hundred square centimeters. The reliability of 3D printed

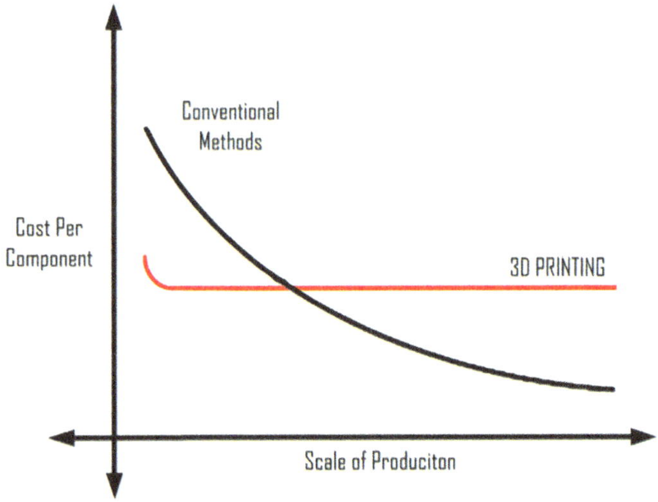

Fig. 2 Cost per component produced as related to the number of components produced (Busachi et al. 2017)

products is another significant concern in their deployment for higher-end procedures, such as mechanical ventilators, where the risks associated with device failure are far more critical as compared to products such as PPEs. This lack of confidence in products is accentuated by the lack of regulatory approvals for the same, which means the absence of a figure of responsibility in the case of adverse failures (Sarkis et al. 2020).

2.2 PPE—Face Shield and Respirators

A necessary consequence of a highly contagious disease is that the primary caregivers responsible for treating infected patients are at a heightened risk of contracting the infection themselves. This necessitates appropriate protective gear for such frontline professionals to avoid the workforce's incapacity throughout the pandemic. Moreover, it is vital to maintain continuous availability of personnel protective equipment or PPE for individuals along the first line of response such as doctors, nurses, etc. However, given the unprecedented nature of the COVID-19 pandemic, these supply chains were quickly overwhelmed by the growing demand for protective gear. The resulting shortages of appropriate PPEs forced hospitals into rationing their supply of such equipment. Therefore, given the uncertain duration of time to follow, alternative sources were identified, and community-sourced 3D printed alternatives were accepted as reasonable choices. Most popularly, face shields used to cover faces with a transparent plastic sheet were developed using additive manufacturing. Masks and respirators have also been designed to be producible via 3D printing.

However, there are numerous challenges associated with the additive manufacturing of such protective gear. One of the foremost concerns with 3D printed PPE arises from the materials used to fabricate them and their compatibility with sterilization procedures employed at medical establishments. Given the limited range of fabrics used to manufacture components additively, there is always uncertainty about the materials' responses to the various sanitization techniques used as medical standards for cleaning protective equipment. While some materials may disintegrate in the presence of chemical disinfectants or UV radiation, others may have a porosity which disallows effective cleaning (Sarkis et al. 2020). Furthermore, the implicit combination of the material and printing method defines the functional properties of the build with significantly different characteristics. For example, it has been observed that the build's strength is primarily influenced by the build orientation, as opposed to other characteristics such as feed rate and layer height (Sandhu et al. 2020). The principal reason behind the deterioration of these quality aspects is the inadequate general guidelines on the authorized usage of additively manufactured products in medical establishments. Especially in terms of PPE, 3D printed products have mostly been designed by non-authorized citizens with community-sourced design files. As a result, these designs may not be granted regulatory approval via standardized demarcations, making it difficult for them to be sold and not given to hospitals in exchange for money (Tarfaoui et al. 2020).

3 Possible Solutions

Although the issues discussed above are highlighted as the growth of usage during a pandemic, their instances in the form of challenges are prevalent in 3D printing while producing medical devices. Therefore, it is also important to note the availability of solutions to such issues, both proposed and primarily accepted. This section aims to discuss a few relevant instances of such solutions in the specific context of the deployment of additive technologies during COVID-19.

3.1 Extensive Adoption of Allied Technologies

To consolidate the advantage of 3D printed products, it is essential to exploit the inherent advantages which make them lucrative replacements for their conventional counterparts. These advantages include design flexibility, customizability, material economy, etc. However, the actualization of such characteristics requires the adoption of numerous allied technologies associated with additive manufacturing, broadly categorized as pre-fabrication and post-fabrication processes.

Allied technologies are processes used before, during, and after the fabrication of a part using the central manufacturing technology. Pre-fabrication methods lead up to production, involved with the design and development of the product to be fabricated.

For a specific instance, to reproduce a component that may already exist, the design of the component in question is generated through reverse engineering methodologies. These include 3D scanning, point cloud interpolation, and CAD modeling of the design itself. Interestingly, the 3D imaging technologies used here, such as laser scanning, CT scanning, and MRI, are also deployed to understand better and better fit the product's design to a specific patient. In the current pandemic context, such technologies have been used to develop effective and comfortable face masks and respirators for frontline professionals such as doctors and nurses. The process flow for the same has been described in Fig. 3. The imaging of the specific user is carried out using one of the available modalities ranging from laser scanning to simple photography. These images are processed using appropriate mathematical tools to generate 3D point clouds, parametrically representative of the user's face, thus providing the required characteristic features of the surface profile. The mask/respirator is to be placed. Following the user comfort, this requires adjusting the design parameters to provide maximum airway sealing and reducing facial lines observed after elongated periods of wearing them (Swennen et al. 2020; Makowski and Okrasa 2019).

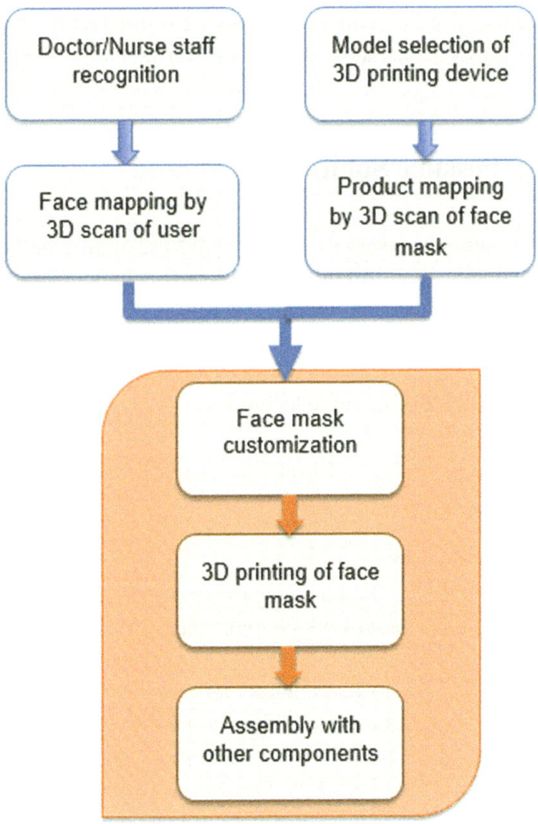

Fig. 3 Customized face mask production flow (Makowski and Okrasa 2019)

However, the mandatory inclusion of the post-processing step for extracting advantages of 3D printing is arguably more of a disadvantage. Issues such as dimensional accuracy and material strength characteristics may be addressed by post-processes such as machining based finishing and thermal-mechanical treatments (HIP, CIP, etc.), which eventually adds overall production cost and time. Statistically, it has been found that only the feed rate has a significant effect on the material removal rate from the finished product, whereas cutting speed and depth of cut are insignificant (Singh et al. 2020). Therefore, there should be a fair trade-off between the design parameters and process-related investments to avoid the incompetency of additive manufacturing over conventional processes.

3.2 Repurposing Over Redesign

In the case of medical devices and equipment, legal issues are a major hurdle between development and marketing and pose a significant risk even after deployment. These issues arise from regulatory approvals, intellectual property rights, usage safety and risk responsibility, standardized testing, and insurances. Therefore, manufacturers and investors tend to remain inertial with established design, development, and production methodologies to minimize risks by eliminating those associated with adopting newer, unconventional methods such as 3D printing. Further, the lack of guidelines related to design adjustments and patient-specific customizations makes the product difficult to adopt flexibility as an advantage. One way to shunt this issue is to use additive methods to repurpose existing products to newer, more relevant, and urgent products. In one instance, Erickson et al. (2020) utilized helmets, used otherwise for arthroplasty procedures, for COVID-19 related treatment procedures by making necessary modifications to the design. In doing so, they exploited the additive manufacturing processes such as SLA and SLS to produce modified manifolds to replace the fan cover, as reproduced in Fig. 4. This modified design was developed in conversation with the original equipment manufacturer, Stryker (2020). The tests were carried out in coordination with university researchers and private consultants,

Fig. 4 Helmet modification component to arthroplasty overall protective gear (Erickson et al. 2020)

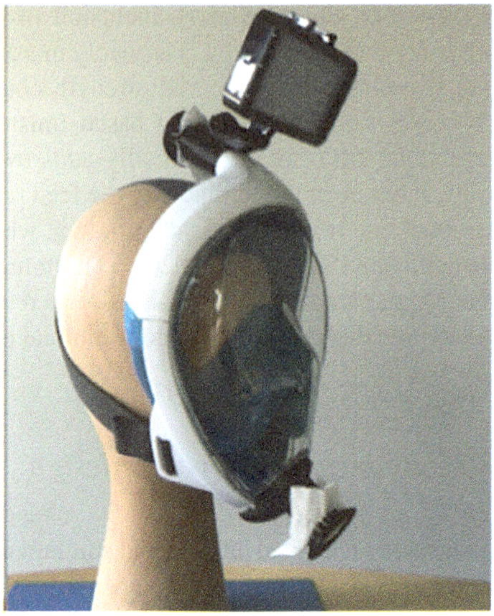

Fig. 5 Diving mask with an upper adaptor (FFP2/N95 mask piece clipped on) alongside an adaptor for lights and a lower adaptor (FFP2/N95 mask piece clipped on) (Thierry et al. 2020)

which established that the product met HEPA standards, and no patient-related risks were identified.

Considering similar instances, components from diver costumes used for underwater exploration have been repurposed to serve as PPE. For example, Thierry et al. (2020) modified a consumer device produced by a sports company for diving and snorkeling to render them usable for frontline medical professionals. By attaching four different kinds of modified disposable components (one clip, one joint, and two adapters), printed using fused deposition modeling as shown in Fig. 5, the consumer market design was altered to the requirements of the treatment procedures. The testing was carried out with the approval of a local hygiene team. The advantage of this mask was to eliminate the necessity of separate FFP2 masks and face shields by providing one scaffold to provide all necessary protection. This also allows for easy communication between doctors and ensures clear visibility.

Such development requires focused work to achieve the essential aspects of the product and enables collaboration with parties who own the intellectual property of the product being developed. Further, the pre-established manifold of the majority of the design allows for readily deploying modified devices compared to the testing and approval time associated with designs created from scratch.

4 Future Opportunities

Additive manufacturing has been prone to numerous challenges that prevent widespread deployment, primarily because of the lack of confidence in involved technologies. The challenges have been persistent, long before the COVID-19 pandemic highlighted them. Given the relative novelty of these processes compared to their machining, casting, or forming based counterparts, these challenges have been accepted as developmental issues requiring incremental evolution and have therefore been able to persist all along. Moreover, the narrow requirements for 3D printing materials make them highly competitive products with very high intellectual property investments, and therefore, carrying the labels of extremely high prices. However, the current scenario has forced developers and manufacturers to take wider cognizance of 3D printing and note the issues that plague the technology. This demands a more competitive yet innovative environment to form with the 3D printing community and solve these problems faster than before. Several upcoming technologies have already shown promising alternatives to address the issues highlighted above. For instance, in a particular work, Walker et al. (2019) described the implementation of an additive manufacturing system that allows for high area rapid printing (HARP), therefore, granting both volume and speed of production. In another instance, Kelly et al. (2017) introduced an apparatus for photopolymerization-based 3D printing, which allows for high resolution given the inherent absence of layers. This ensures a system of high-grade finish with reduced time and cost associated with post-production processes. The research and development of cost-effective materials such as Poly-lactic-acid (PLA) are more promising to address these issues Singh et al. (2019). PLA has been successfully employed to develop medical devices to share direct contact with tissues and organs. Singh et al. (2019) presented a case study where PLA structures, embedded with waste natural fiber, are printed using a fused deposition approach. In another study by Ploch et al. (2016), the incapability of printing flexible and soft tissue parts is resolved by a surrogate gelatin material, with 3D printing and casting process unification. Finally, there is an emergent need for large scale technological collaborations with developers from diverse technical, economic, and social backgrounds to increase its acceptance across larger markets. While additive manufacturing has already found a niche within the medical industry, it is yet to be established as a conventional standard for manufacturing mainstream medical devices.

5 Conclusions

In this work, several challenges associated with implementing 3D printed medical products, such as PPEs and ventilators during the COVID-19 pandemic, have been discussed. As observed in the literature, solutions to these challenges have been elaborated after carefully reviewing the accompanying factors for the initial deployment of 3D printed products during the pandemic. The primary sources of problems are a limited range of materials and a lack of inefficient and slow regulatory approvals for new 3D printed products. Moreover, the minor issue regarding the durability of the 3D printed products has also been observed. Through this discussion, it has been evident that the most guaranteed way towards broader acceptance of additive manufacturing in the medical industry is through accelerated research and learnings from the limitations observed through the COVID-19 pandemic. Finally, the reviewed solutions in this work could be effectively extended as future opportunities within various forms of additive manufacturing. This work will help enthusiastic researchers and doctors in the domain of 3D printing to address the emergent medical needs for pandemics like COVID-19.

Acknowledgements The authors would like to acknowledge the support received from Mechatronics and Robotics Laboratory, IIT Guwahati. Moreover, the authors are also grateful to all the technical and non-technical communities for continuous monitoring of the issues raised during the COVID-19 pandemic.

References

Attaran M (2020) 3D printing role in filling the critical gap in the medical supply chain during COVID-19 pandemic. Am J Ind Bus Manag 10(05):988

Berman B (2012) 3-D printing: the new industrial revolution. Bus Horiz 55(2):155–162

Busachi A, Erkoyuncu J, Colegrove P, Martina F, Watts C, Drake R (2017) A review of additive manufacturing technology and cost estimation techniques for the defence sector. CIRP J Manuf Sci Technol 19:117–128

Callaway E, Cyranoski D, Mallapaty S, Stoye E, Tollefson J (2020) The coronavirus pandemic in five powerful charts. Nature 579(7800):482–483

Choong YYC, Tan HW, Patel DC, Choong WTN, Chen CH, Low HY, Tan MJ, Patel CD, Chua CK (2020) The global rise of 3D printing during the COVID-19 pandemic. Nat Rev Mater 5:637–639

Erickson MM, Richardson ES, Hernandez NM, Bobbert DW II, Gall K, Fearis P (2020) Helmet modification to ppe with 3d printing during the COVID-19 pandemic at duke university medical center: a novel technique. J Arthroplasty 35(7S):S23–S27

Gibson I, Rosen DW, Stucker B (2014) Additive manufacturing technologies. Springer, New York. https://doi.org/10.1007/978-1-4939-2113-3

Goyal P, Choi JJ, Pinheiro LC, Schenck EJ, Chen R, Jabri A, Satlin MJ, Campion TR Jr, Nahid M, Ringel JB, Hoffman KL (2020) Clinical characteristics of COVID-19 in New York city. N Engl J Med 382(24):2372–2374

Higgins R, Martin E, Vesperi MD (2020) An anthropology of the Covid-19 pandemic. Anthropolo Now 12(1):2–6

Ishack S, Lipner SR (2020) Applications of 3D printing technology to address COVID-19 related supply shortages. Am J Med 133(7):771–773

Ivanov D, Dolgui A (2020) Viability of intertwined supply networks: extending the supply chain resilience angles towards survivability. A position paper motivated by COVID-19 outbreak. Int J Prod Res 58(10):2904–2915

Kelly B, Bhattacharya I, Shusteff M, Panas RM, Taylor HK, Spadaccini CM (2017) Computed axial lithography (CAL): Toward single step 3D printing of arbitrary geometries. arXiv preprint arXiv: 1705.05893

Kenney ME (2013) Cost reduction through the use of additive manufacturing (3D Printing) and collaborative product lifecycle management technologies to enhance the Navy's Maintenance Programs (No. NPS-LM-13-087). Naval Postgraduate School Monterey Ca Graduate School of operational and information sciences

Liaw CY, Guvendiren M (2017) Current and emerging applications of 3D printing in medicine. Biofabrication 9(2):024102

Makowski K, Okrasa M (2019) Application of 3D scanning and 3D printing for designing and fabricating customized half-mask facepieces: a pilot study. Work 63(1):125–135

Manero A, Smith P, Koontz A, Dombrowski M, Sparkman J, Courbin D, Chi A (2020) Leveraging 3D printing capacity in times of crisis: recommendations for COVID-19 distributed manufacturing for medical equipment rapid response. Int J Environ Res Public Health 17(13):4634

McKibbin WJ, Fernando R (2020) The global macroeconomic impacts of COVID-19: seven scenarios. CAMA Working Paper No. 19/2020. https://dx.doi.org/10.2139/ssrn.3547729

MIT News Office (2011) Printing off the paper: MIT research continues to push the boundaries of the burgeoning technology of 3-D printing. MIT News. https://news.mit.edu/2011/3d-printing-0914

Moein ST, Hashemian SM, Mansourafshar B, Khorram-Tousi A, Tabarsi P, Doty RL (2020) Smell dysfunction: a biomarker for COVID-19. Int Forum Allergy Rhinol 10:944–950

Mollenkopf DA, Ozanne LK, Stolze HJ (2020) A transformative supply chain response to COVID-19. J Serv Manag. https://doi.org/10.1108/JOSM-05-2020-0143

WHO. Newsroom (2020) COVID-19 significantly impacts health services for noncommunicable diseases. https://www.who.int/news-room/, Accessed at Aug 2020

Nikolova MP, Chavali MS (2019) Recent advances in biomaterials for 3D scaffolds: a review. Bioact Mater 4:271–292

Novak JI, Loy J (2020) A critical review of initial 3D printed products responding to COVID-19 health and supply chain challenges. Emerald Open Res 2(24):24

Ploch CC, Mansi CS, Jayamohan J, Kuhl E (2016) Using 3D printing to create personalized brain models for neurosurgical training and preoperative planning. World Neurosurg 90:668–674

Rengier F, Mehndiratta A, Von Tengg-Kobligk H, Zechmann CM, Unterhinninghofen R, Kauczor HU, Giesel FL (2010) 3D printing based on imaging data: review of medical applications. Int J Comput Assist Radiol Surg 5(4):335–341

Sandhu K, Singh JP, Singh S (2020) Some investigations on the tensile strength of additively manufactured polylactic acid components. In: Singh S, Prakash C, Ramakrishna S, Krolczyk G (eds) Advances in materials processing (Lecture notes in mechanical engineering). Springer, Singapore. https://doi.org/10.1007/978-981-15-4748-5_22

Sarkis J, Cohen MJ, Dewick P, Schröder P (2020) A brave new world: lessons from the COVID-19 pandemic for transitioning to sustainable supply and production. Resour Conserv Recycl 159:104894

Shahrubudin N, Koshy P, Alipal J, Kadir MHA, Lee TC (2020) Challenges of 3D printing technology for manufacturing biomedical products: a case study of Malaysian manufacturing firms. Heliyon 6(4):e03734

Singh S, Prakash C, Singh M, Mann GS, Gupta MK, Singh R, Ramakrishna S (2019) Poly-lactic-acid: potential material for bio-printing applications. In: Prakash C et al (eds) Biomanufacturing. Springer, Cham. https://doi.org/10.1007/978-3-030-13951-3_3

Singh S, Singh G, Sandhu K, Prakash C, Singh R (2020) Investigating the optimum parametric setting for MRR of expandable polystyrene machined with 3D printed end mill tool. Mater Today: Proc 33(3):1513–1517

Stryker (2020) Medical surgical equipment. https://www.stryker.com/in/en/portfolios/i/medical-surgical-equipment.html

Swennen GR, Pottel L, Haers PE (2020) Custom-made 3D-printed face masks in case of pandemic crisis situations with a lack of commercially available FFP2/3 masks. Int J Oral Maxillofac Surg 49(5):673–677

Tarfaoui M, Nachtane M, Goda I, Qureshi Y, Benyahia H (2020) 3D printing to support the shortage in personal protective equipment caused by COVID-19 pandemic. Materials 13(15):3339

Thierry B, Célérier C, Simon F, Lacroix C, Khonsari RH (2020) How and why use the EasyBreath® Decathlon surface snorkeling mask as a personal protective equipment during the COVID-19 pandemic? Eur Ann Otorhinolaryngol Head Neck Dis 137(4):329–331

Tino R, Moore R, Antoline S, Ravi P, Wake N, Ionita CN, Morris JM, Decker SJ, Sheikh A, Rybicki FJ, Chepelev LL (2020) COVID-19 and the role of 3D printing in medicine. Print Med 6(11):1–8

Truog RD, Mitchell C, Daley GQ (2020) The toughest triage—allocating ventilators in a pandemic. N Engl J Med 382(21):1973–1975

Ventola CL (2014) Medical applications for 3D printing: current and projected uses. Pharm Ther 39(10):704–711

Walker DA, Hedrick JL, Mirkin CA (2019) Rapid, large-volume, thermally controlled 3D printing using a mobile liquid interface. Science 366(6463):360–364

WHO coronavirus disease (COVID-19) dashboard (2020) Geneva: World Health Organization. https://covid19.who.int/

Covid-19 Success Stories of 3D Printing

Harmanpreet Singh, Sagarika Bhattacharjee, and Puneet Bawa

Abstract The manufacturing sector got revamped with the introduction of 3D printing technology. Additive manufacturing is in great demand because the products made by this technology require minimal post-processing, that too in the case of applications requiring high finishing. 3D printing is capable of producing complex shapes, as the technology uses layer by layer processing. Initially, the technology was developed to make the physical prototypes rapidly for an approximation of the real product, but now the real products are made using various 3D printing techniques which are classified according to the powder, solid, and liquid-based forms. The recent outbreak of the Covid-19 pandemic, from the Hubei province of China, has challenged the pharmaceutical and related medical industries with the need arising for the various medical equipment and gadgets protecting the spread of the virus. Researchers from around the world suggested various solutions for tackling the problem, introducing new products that required instant manufacturing. 3D printing is a reliable technology that succored many research groups and industries to develop and produce the products which were proved to be vital in these hard times. The chapter talks about the various success stories of the researchers who fabricated various products using 3D printing technology to tackle the complications of the Covid-19 pandemic.

Keywords 3D printing · Pandemic · Personal protective equipment (PPE) · COVID 19 · Sustainability

---

H. Singh (✉)
Department of Mechanical Engineering, Thapar Institute of Engineering and Technology, Patiala 147004, Punjab, India

S. Bhattacharjee
Department of Physics and Nanotechnology, SRM Institute of Science and Technology, Kattankulathur 603203, Tamil Nadu, India

P. Bawa
Chitkara University Institute of Engineering& Technology, Centre of Excellence for Speech and Multimodal Laboratory, Chitkara University, Rajpura 140401, Punjab, India

K. Sandhu et al. (eds.), *Emerging Applications of 3D Printing During CoVID 19 Pandemic*, Lecture Notes in Bioengineering, https://doi.org/10.1007/978-981-33-6703-6_11

1 Introduction

The SARS-CoV-2 (COVID 19) initiated in December of 2019 in Wuhan, Hubei, China, which rapidly outbreak globally and was declared a pandemic by WHO. The illness due to the virus ranges from asymptomatic to severe, where the transmission is from person to person usually by close contact (Francois et al. 2020; https://www.elsevier.com/__data/assets/pdf_file/0010/977698/novel-coronavirus-covid-19-infection-2020-07-28.pdf). The detailed scientific knowledge about the virus was incomplete, so the primitive step of precaution was social distancing and treatment included the 14–15 days quarantine along with adequate oxygenation and medical support in severe cases (https://www.elsevier.com/__data/assets/pdf_file/0010/977698/novel-coronavirus-covid-19-infection-2020-07-28.pdf). The outbreak of the virus halted the activities globally, causing uncertainty and fear leading to anxiety, related disorders (https://www.elsevier.com/connect/coronavirus-information-center?dgcid=_SD_banner#clinical-information; https://covid-19.elsevier.health/en-US/resources/mental-health). There was continuous elevating demand in the personal protective equipment and other non-contact devices helpful in preventing the transmission which was produced and supplied to the containment zones (Tino et al. 2020).

Additive manufacturing (AM) is the new way of manufacturing products and is gaining popularity exponentially in various domains. Additive manufacturing is an easy way of designing and producing products. The designer uses the software package to design the required product and that design is converted to suitable formats like (.stl) which are read by the machine and the required responses are generated for production. The starting materials used in the AM process can be of different categories like solid, liquid, paste, and sheet, and based on this there are few types of AM techniques and machine setups. During the pandemic period, the AM has played an important role in producing some new and also a few existing products. A few of these success stories will be discussed further in the chapter.

2 Outbreak of COVID-19

The COVID-19 was identified in Wuhan China, in December 2019, and during March 2020, it was considered a pandemic (COVID-19 dashboard by the center for systems science and engineering (CSSE) at John Hopkins University (JHU)). Though the effect of the pandemic was observed worldwide, few nations in different continents were affected the most. The comparison of few countries from different continents which were affected the most as of 17 April 2020 is depicted in the graph shown in Fig. 1, COVID-19 was at peaks in the USA, while South Africa was least affected among different countries depicted in the graph (Coronavirus disease 2019 situation report-88 2020). The impact of the disease worldwide made the researchers use the technology for solving the present needs for the essential products and also some

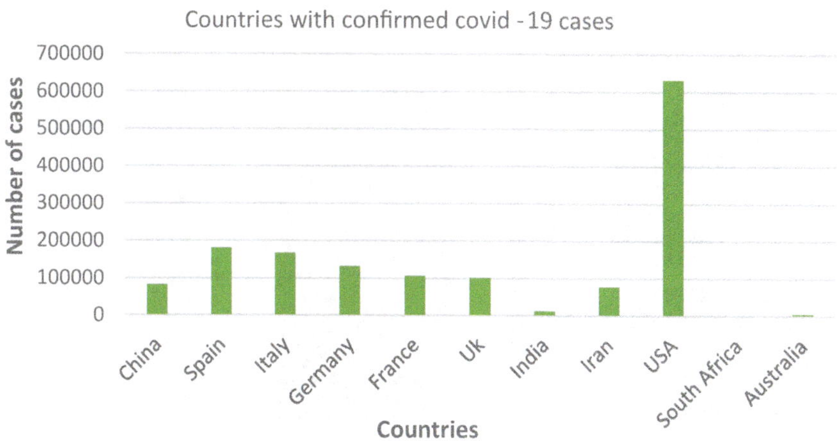

Fig. 1 Graph for the confirmed cases of COVID-19 as of 17 April 2020 (Coronavirus disease 2019)

products were made on the conditions aroused, by making the use of 3D printing. The internet served as the best tool for the production using 3D printing by sharing the designs of the product through the open-source platform to fight against these tough times. A few of the successful products from different nations are discussed in a further section.

3 Covid-19 Success Stories of 3D Printing

Soon after the declaration of the COVID-19 pandemic, researchers around the world started investigating various ways to tackle the virus. The research was not just focused on testing and treating the patients, but also included the new ways to tackle the situations without further contamination as the virus was spread by community transmission (How COVID-19 Spreads. U.S. Centers for Disease Control and Prevention (CDC). 18 Sep 2020). Various researches were done to make the products to deal with the COVID-19 situation. 3D printing played a great role in making new products faster and efficiently (Amin et al. 2020; Erickson et al. 2020; Flangan and Ballard 2020; Ms et al. 2020).

3.1 Door Handles and Push Buttons

Recent researches revealed that the COVID-19 virus can survive for 72 h on surfaces made of plastic and stainless steel (Thornburg 2020). Due to this, there were high chances of contamination through the door handles and the button of the elevators,

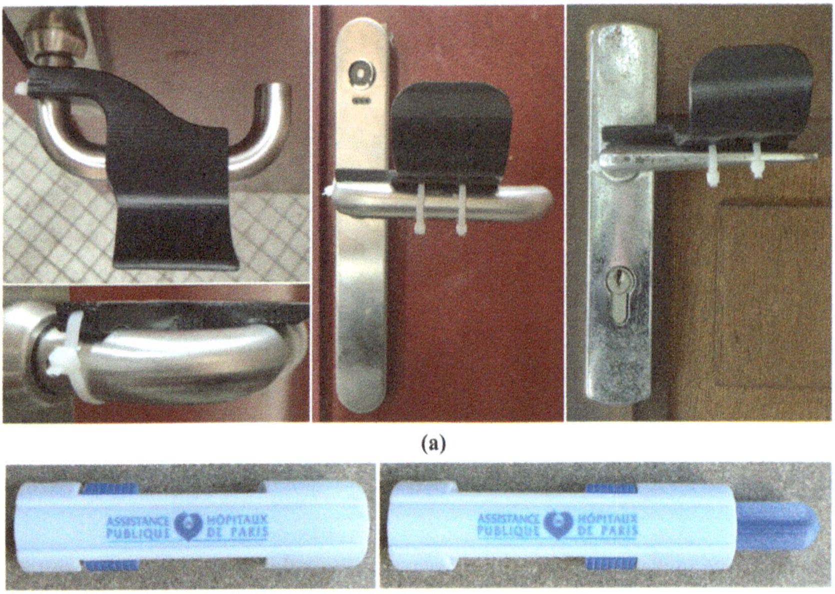

(a)

(b)

Fig. 2 3D printed **a** Hand free door openers. **b** Push buttons in closed (left) and open position (right) (Francois et al. 2020)

etc. Francois et al. (Francois et al. 2020) from France focused on this problem and made the hands-free door openers and the push buttons as shown in Fig. 2a, b through the use of additive manufacturing techniques in a short period and delivered it to the hospitals and health centers. Fused deposition modeling technique was used to make the products, wherein the raw material used was Acrylonitrile Butadiene Styrene (ABS) in black and white color. The door openers made were of two kinds based on their installation, the one of which was fixed using the help of the clips and the other one was attached using the three cables. With the good collaboration of machines and people, the products were supplied smoothly.

The next relevant need of an hour item for the current situation was the Personal Protective Equipment (PPE). The frontline workers handling the patients are at the most risk of getting infected if they are without the proper safety gear (Covid-19 FAQ's (2020)). Mask alone doesn't serve the purpose as the add-on safety needs to be implemented for the frontline workers in the hospitals. Personal protective equipment covers all protection items used like masks, face shields, gloves, isolation gowns, etc. 3D printing has played a good role in developing a few of the PPE's in this tough situation.

3.2 Face Shields

This is the most important add-on device to the mask and goggles that help in preventing the infection during close contact with the patients suffering from the COVID-19. The shield ae easy to clean and can be reused multiple times before disposal.

Armijo et al. from the US have produced the face shields using 3 D printing and other available products. The shield depicted in Fig. 3 was made to avoid the infection and also the cleaning protocol was set so that it can be reused multiple times. FDM technique of 3D printing was utilized to make the 112 face shields in approximately 72 h. Similarly, the method was replicated by other military squadrons to produce the 100 face shields in 72 h to be delivered to the Tripler Army medical center. The basic criteria of 3D printing were followed by making the design using the CAD software followed by the slicing and transferring the.stl file formats to the printer for making the required products. The group made the solid headband and the chin piece to be used in making the face shield using the 3D printing technique, rest of the other materials were purchased commercially which included the transparent faceguard and other supporting materials. The 3D printing can serve the best emergency purpose of the production of unique and essential items on a small scale which otherwise are available on a limited basis, provided that one should have a good command over geometric modeling techniques (Kantor 2020).

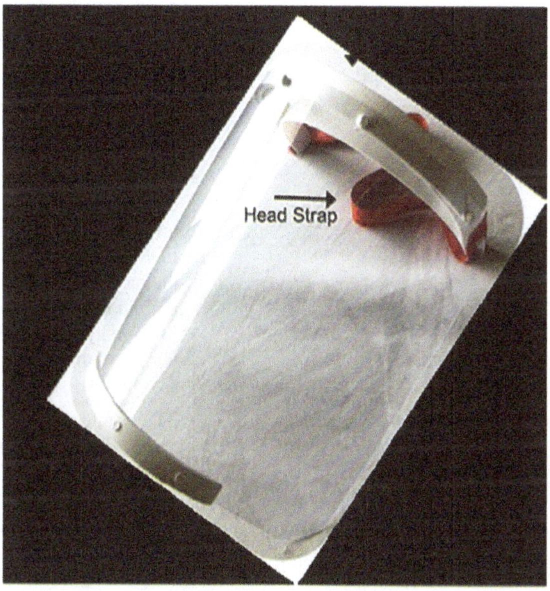

Fig. 3 Reusable face shield made using 3D printing Armijo et al. (2021)

A similar kind of work was performed by a few more researchers and a brief is discussed in Table 1 mentioning the techniques used.

Worldwide the different approaches were adopted to serve the current need for the face shield as it was helpful in protecting the masks and overall infection through the droplets. The face shields can be reused by following the simple procedures of cleaning them.

Table 1 List of researchers who have produced the face shield using 3D printing

S. No	Author	Country/region	Methodology
1.	Wesemann et al. (2020)	Germany	• FDM technique was used, due to minimal post-processing requirements • Four kinds of open-source STL were selected and produced for analysis • Easy 3D face shield design was most effective in terms of printability, assembly, and protection • For large-scale production, the RC2 face shield design was found more effective as it has a stacked dataset
2.	Amin et al. (2020)	US	• Airwolf 3D STL file was selected for face shield production • FDM printer was used for making the face shields • Super Sani cloth germicidal wipes are recommended for cleaning purposes of the face shield
3.	Gomes et al. (2020)	Brazil	• Three different models of face shields were made based on the STL design available as open source • RC2 model was found to be the best based on the protection and comfort level • FDM printer was used for making the face shield headband with ABS premium MG94 material • Sodium hypochlorite (0.1%) or 68–71% ethanol was used for 1-min soaking to disinfect the face shield
4.	Novak et al. (2020)	Australia	• 37 separate designs for face shields and 31 designs for face masks were collected • FDM printer was used for making the face shields and face masks • Face shield takes almost half time for printing as the face masks and also the material used in the shield is also less compared to masks

3.3 Masks

This is the most important and demanded product to get protected against the COVID-19 virus. Due to the outbreak of the pandemic, there has been a huge change in mask consumption, which lead to a shortage of standardized N95 and FFP2/3masks in the world (Ranney et al. 2020; Wu et al. 2020; Swennen et al. 2020). Swennen et al. (2020) from the UK used the selective laser sintering process for making the custom-made mask shown in Fig. 4. The mask can be sustainable in terms of reusability as two of the components are reusable viz a face mask and a filter membrane which requires disinfection by some solution soaking. While some components need to be replaced as they are not fit for disinfecting. These components are easily available in the market and can be assembled to form a complete custom-made mask. Thus, this method serves as an alternative to available methods for producing the standard masks in this pandemic situation. Apart from this, several other designs were proposed for the printing of masks than standard techniques (Tino et al. 2020). In addition to the printing of the mask, various other add-on products were made using the 3D printing techniques which enhances the protection while using the mask. The snorkel mask adapter was among these products that can be used in mask shortage situations, however, the replacement with the standard N95 mask is strictly not suggested (Dalla et al. 2020). The mask adapter is an add-on component that can be attached to the mask, which serves the same standard guidelines as the N95 mask and can help in long term usage of the mask in these tough times where the production, supply chain is the major issue (Imbrie-Moore et al. 2020). Recyclability or reuse of the mask is a big problem, as that causes a huge amount of solid waste. He et al. (2020) from Hungary developed the biodegradable mask filters produced by the combination of

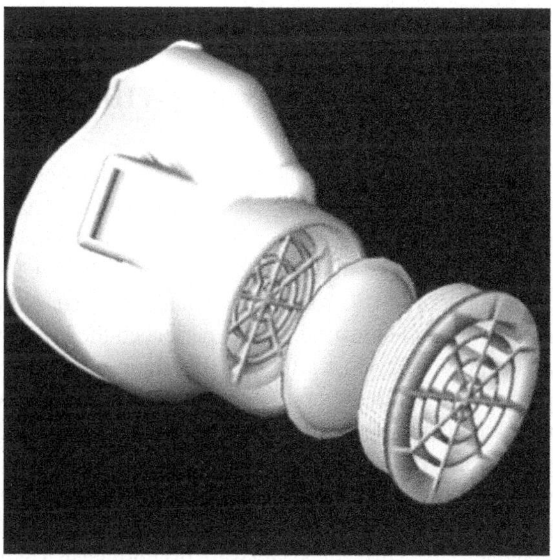

Fig. 4 Image of the prototype mask made by 3D printing technique (Swennen et al. 2020)

electrospinning and 3D printing. These filters serve to be more efficient in terms of sustainability as using them the user need to change the whole mask, while the filter can be replaced.

3.4 Ventilator Accessories

Ventilator as a whole is the machine required for critical patients suffering from the COVID-19. In this pandemic situation, there is a great shortage of ventilators in hospitals. 3D printing can make certain accessories including valves that are the essential components of the ventilators (Tino et al. 2020; Arora et al. 2021; Challenges and solutions in meeting up the urgent requirement of ventilators for COVID-19 patients 2020).

3.5 Few Other Products (Arora et al. 2021)

A variety of other products can be made using 3D printing that find their suitable application in this pandemic period. Some products are specially designed in this period to tackle the problems related to contamination. Hand sanitization is a common solution to prevent the virus when we come in contact with other people. Workplaces have provided the facilities of hand sanitization where the sanitizer holders were made in use, as they prevent direct touching and can be operated without touching the sanitizer bottle. Similarly, the oxygen valves were designed and made using 3D printing in Italy.

4 Supply Chain Issues in COVID-19

The regular supply of materials from the industries to the retail shops in a smoother way can be called the supply chain of finished materials. The halts in the supply chain is not a new condition and were also observed earlier due to natural calamities (Chopra and Sodhi 2014; Choi et al. 2020; Budd et al. 2011; Canis 2011; Chongvilaivan 2012; Shaheen et al. 2017; Kumar and Chandra 2010). Now due to the pandemic, major economies and industrial hubs experienced the lockdowns due to which the demand and supply were most affected. 3D printing played a great role in making the new products to tackle the problems and also making the other possible way out for the essential components like face shields, masks, and also some accessories of the ventilators. The most helpful part was that the tested designs were uploaded on the open-source websites from where they were made in use by transferring it to the 3D printers and making the required products (Salmi et al. 2020). Some PPE's were such that they can be used once and were in shortage in a pandemic situation, for which the various other add-on products were made using the 3D printing techniques, which helped in enhancing the life cycle of the PPE's and also contributing towards the sustainability (Imbrie-Moore et al. 2020; Rowan and Laffey 2020).

5 Conclusion

Worldwide the people have faced the consequences of the COVID-19 pandemic, the cases are still rising globally. 3D printing was the most promising solution for the production of the necessary items in demand like PPE's and also some new solutions like door handles and push buttons which halted the virus from further spread. Researchers from around the globe were successful in making use of 3D printing technology for making the products. The standard designs were uploaded on the open-source websites, which were downloaded at different locations and used to feed the.stl files to the printers for production. This technique also helped in running the global supply chain adjourn by giving alternate solutions. The 3D printing technology can be improved in terms of the production time as currently it is not suitable for large-scale production and also some research into the new materials can make it in future demands.

References

Amin D, Nguyen N, Roser SM, Abramowicz S (2020) 3D Printing of face shields during COVID-19 pandemic: a technical note. J Oral Maxillofac Surg. https://doi.org/10.1016/j.joms.2020.04.040

Armijo PR, Markin NW, Nguyen S, Ho DH, Horseman TS, Lisco SJ, Schiller AM (2021) 3D printing of face shields to meet the immediate need for PPE in an anesthesiology department during the COVID-19 pandemic. Am J Infect Control. https://doi.org/10.1016/j.ajic.2020.07.037

Arora PK, Arora R, Haleem A, Kumar H (2021) Application of additive manufacturing in challenges posed by COVID-19. Mater Today: Proc. https://doi.org/10.1016/j.matpr.2020.08.323

Budd L, Griggs S, Howarth D, Ison S (2011) A fiasco of volcanic proportions? Eyjafjallajokull and the closure of European airspace. Mobilities 6:31–40

Canis B (2011) Motor vehicle supply chain: effects of the Japanese earthquake and tsunami. Congressional Research Service Publishing: Washington, DC, USA, pp 1–16

Challenges and solutions in meeting up the urgent requirement of ventilators for COVID-19 patients. Diabetes Metab Syndr: Clin Res Rev (2020). https://doi.org/10.1016/j.dsx.2020.04.048

Choi TY, Rogers D, Vakil B (2020) Coronavirus is a wake-up call for supply chain management. Harv Bus Rev

Chongvilaivan A (2012) Thailand's 2011 flooding: Its impact on direct exports and global supply chains. ARTNeT Policy Brief 113:1–34

Chopra S, Sodhi M (2014) Reducing the risk of supply chain disruptions. MIT Sloan Manag Rev 55:72–80

Coronavirus disease 2019 situation report-88 (2020). https://www.who.int/docs/default-source/coronaviruse/situation-reports/20200417-sitrep-88-covid-191b6cccd94f8b4f219377bff55719a6ed.pdf?sfvrsn=ebe78315_6

Covid-19 FAQ's. https://www.asahq.org/about-asa/governance-and-committees/asa-committees/committee-on-occupational-health/coronavirus/clinical-faqs. Accessed 20 Aug 2020

COVID-19 dashboard by the center for systems science and engineering (CSSE) at John Hopkins University (JHU). ArcGIS. John Hopkins University

Dalla S, Shinde R, Ayres J, Waller S, Nachtigal J (2020) 3D printed Snorkel Mask adapter for failed N95 Fit tests and PPE shortages. medRxiv. https://doi.org/10.1101/2020.08.18.20174540

Erickson MM, Richardson ES, Hernandez NM, Bobbert DW, Gall K, Fearis P (2020) Helmet modification to PPE with 3D printing during the COVID-19 Pandemic at Duke University Medical Centre: a novel technique. J Arthroplasty. https://doi.org/10.1016/j.arth.2020.04.035

Flangan ST, Ballard DH (2020) 3D Printed face shields: a community response to the COVID-19 global pandemic. Academic Radiology. https://doi.org/10.1016/j.acra.2020.04.020

Francois PM, Bonnet X, Kosior J, Adam J, Khonsari RH (2020) 3D-printed contact-free devices designed and dispatched against the COVID-19 pandemic: the 3D COVID initiative. J Stomatol Oral Maxillofac Surg. https://doi.org/10.1016/j.jormas.2020.06.010

Gomes BDA, Queiroz FLC, Pereira PLDO, Barbosa TV, Tramontana MB, Afonso FAC, Garcia EDS, Borba AM (2020) In-house three-dimensional printing workflow for face shield during COVID-19 pandemic. J Craniofac Surg. https://doi.org/10.1097/SCS.0000000000006723

He H, Gao M, Illes B, Molnar K (2020) 3D printed and electrospun, transparent, hierarchical polylactic acid mask nanoporous filter. Int J Bioprinting. https://doi.org/10.18063/ijb.v6i4.278

How COVID-19 Spreads. U.S. Centers for Disease Control and Prevention (CDC). 18 Sep 2020. Archived from the original on 19 Sep 2020. Accessed 20, Sep 2020

https://www.elsevier.com/__data/assets/pdf_file/0010/977698/novel-coronavirus-covid-19-infection-2020-07-28.pdf

https://www.elsevier.com/connect/coronavirus-information-center?dgcid=_SD_banner#clinical-information

https://covid-19.elsevier.health/en-US/resources/mental-health

Imbrie-Moore AM, Park MH, Zhu Y, Paulsen MJ, Wang H and Woo YJ (2020) Quadrupling the N95 supply during the COVID-19 crisis with an innovative 3D-printed mask adaptor. Healthcare. https://doi.org/10.3390/healthcare8030225

Kantor J (2020) Behavioral considerations and impact on personal protective equipment use: early lessons from the coronavirus (COVID-19) pandemic. J Am Acad Dermatol 1087–1088

Kumar S, Chandra C (2010) Supply chain disruption by avian flu pandemic for US companies: A case study. Transp J 49:61–73

Ms DP, Rao YJ, Mitic K, Obaid SN, Pierce D, Huckenpahler J, Berger J, Goyal S, Loew MH (2020) Rapid prototyping of reusable 3D-printed N95 equivalent respirators at the George Washington University. http://dx.doi.org/10.20944/preprints202003.0444.v1

Novak JI, Loy J (2020) A quantitative analysis of 3D printed face shields and masks during COVID-19. Emerald Open Res. https://doi.org/10.35241/emeraldopenres.13815.1

Ranney ML, Griffeth V, Jha AK (2020) Critical supply shortages—the need for ventilators and ventilators and personal protective equipment during the COVID-19 pandemic. N Engl J Med. https://doi.org/10.1056/nejmp2006141

Rowan NJ, Laffey JG (2020) Challenges and solutions for addressing critical shortage of supply chain for personal and protective equipment (PPE) arising from Coronavirus disease (COVID19) pandemic—case study from the Republic of Ireland. Sci Total Environ. https://doi.org/10.1016/j.scitotenv.2020.138532

Salmi M, Akmal JS, Pei E, Wolff J, Jaribion A, Khajavi SH (2020) 3Dprinting in COVID-19: productivity estimation of the most promising open source solutions in emergency situations. Appl Sci. https://doi.org/10.3390/app10114004

Shaheen I, Azadegan A, Lucianetti L, Qi L (2017) Leading organizations through supply chain disruptions: An exploratory study of necessary traits. Rutgers Bus Rev 2:322–337

Swennen GRJ, Pottel L, Haers PE (2020) Custom-made 3D-printed face masks in case of pandemic crisis situations with a lack of commercially available FFFP2/3 masks. Int J Oral Maxillofac Surg. https://doi.org/10.1016/j.ijom.2020.03.015

Tino R, Moore R, Atoline S, Ravi P, Wake V, Ionita CN, Morris JM, Decker SJ, Sheikh A, Rbicki FJ, Chepelev LL (2020) COVID-19 and the role of 3D printing in medicine. 3D Print Med. https://doi.org/10.1186/s41205-020-00064-7

Van Doremalen N, Bushmaker T, Morris DH, Holbrook MG, Gamble A, Williamson BN, Tamin A, Harcourt JL, Thornburg, NJ, Gerber AI, Lloyd Smith JO, Wit ED, Munster VJ (2020) Aerosol and surface stability of SARS-Cov-2 as compared with SARS-Cov-1. N Engl J Med. http://dx.doi.org/10.1056/NEjMc20049973

Wesemann C, Pieralli S, fretwurst T, Nold J, Nelson K, Schmelzeisen R, Hellwig E, Spies BC (2020) 3-D printed protective equipment during COVID-19 pandemic. Materials. https://doi.org/10.3390/ma13081997

Wu HL, Huang J, Zhang CJP, He Z, Ming WK (2020) Facemask shortage and the novel coronavirus disease (COVID-19) outbreak: reflections on public health measures. E Clin Med. https://doi.org/10.1016/j.eclinm.2020.100329

The manufacturer's authorised representative in the EU is Springer Nature Customer Service Centre GmbH, Europaplatz 3, 69115 Heidelberg, Germany. If you have any concerns regarding our products, please contact ProductSafety@springernature.com

Printed and bound by CPI Group (UK) Ltd, Croydon, CR0 4YY
30/09/2026
02225295-0002